Tourism Planning and Development in the Middle East

Tourism Planning and Development in the Middle East

Edited by

Stella Kladou,

Konstantinos Andriotis,

Anna Farmaki

and

Dimitrios Stylidis

CABI

CABI is a trading name of CAB International

CABI
Nosworthy Way
Wallingford
Oxfordshire OX10 8DE
UK

Tel: +44 (0)1491 832111
E-mail: info@cabi.org
Website: www.cabi.org

CABI
200 Portland Street
Boston
MA 02114
USA

T: +1 (617)682-9015
E-mail: cabi-nao@cabi.org

A catalogue record for this book is available from the British Library, London, UK.

<< Insert Library of Congress heading and text here >>

ISBN-13: 9781789249125 (hardback)
9781789249132 (ePDF)
9781789249149 (ePub)

DOI: 10.1079/9781789249149.0000

Commissioning Editor: Claire Parfitt
Editorial Assistant: Lauren Davies
Production Editor: Marta Patiño

Typeset by Exeter Premedia Services Pvt Ltd, Chennai, India
Printed and bound by CPI Group (UK) Ltd, Croydon, CR0 4YY

Contents

About the Editors

Stella Kladou
Mediterranean University, Greece
Email: stellakladou@hmu.gr

Dr Stella Kladou is a postgraduate researcher at the Hellenic Mediterranean University in Crete, Greece. She has served as an Assistant Professor in the Tourism Administration Department at Boğaziçi University in Istanbul, Turkey, and a Senior Lecturer at Sheffield Hallam University in the UK. Her PhD in 'Evaluating Cultural City Brands: The Case of Athens, Istanbul and Rome', was realized in cooperation with the Hellenic Open University in Greece, Istanbul Bilgi University in Turkey and the University of Rome La Sapienza in Italy, and was partly funded by the Italian Ministry of Foreign Affairs and the Scientific and Technological Research Council of Turkey. She has published in prominent academic journals and books of international publishers, and presented at well-esteemed international conferences. She has co-edited books relating to tourism matters, and participated in and/or led several international and national projects as a researcher or consultant. She is an International Place Brand Specialist for the Place Brand Observer. Stella Kladou is on the scientific board of international conferences, the editorial board of the *Journal of Qualitative Research in Tourism*, and the editorial review board of *Annals of Tourism Research*. Her research focuses mainly on place branding, cultural and wine tourism, and digital marketing.

Konstantinos Andriotis
Middlesex University London, UK
Email: k.andriotis@mdx.ac.uk

Dr Konstantinos Andriotis is a Professor in Tourism at Middlesex University London, and in the past he was an Associate Professor and Head of the Hotel and Tourism Management Department at the Cyprus University of Technology. He holds an MSc in International Hospitality Management (University of Strathclyde, 1995), a PhD in Tourism Development and Planning (Bournemouth University, 2000) and a postdoctoral fellowship in Tourism Marketing (Hellenic Open University, 2006). He edits the *International Journal of Tourism Policy*, the *Journal of Qualitative Research in Tourism* and the CABI Regional Tourism Series. He is on the editorial boards of more than 20 journals. He is author of 7 books and over 50 peer-reviewed articles and book chapters. His work has appeared in leading academic journals, including *Annals of Tourism Research, Journal of Travel Research, Tourism Management* and *Journal of Sustainable Tourism*. He has an h-index of 44 and more than 5000 citations. He was listed in the Stanford University list of top 2% of scientists in the world. He has acted as chairman and member of the organizing and scientific committees of various international conferences and has given keynote speeches at several conferences. His research interests include tourism development and planning, alternative forms of tourism, tourism experience and degrowth.

Anna Farmaki
Cyprus University of Technology, Cyprus
Email: anna.farmaki@cut.ac.cy

Anna Farmaki is an Assistant Professor in Tourism Management at the Cyprus University of Technology, Cyprus. She holds a doctorate (PhD) in marketing. She has worked for several years in private institutions of tertiary education in Cyprus and was course leader of the undergraduate hospitality and tourism management programme at the University of Central Lancashire (Cyprus). She has published extensively in reputable, peer-reviewed academic journals and has presented her work at various international conferences, seminars and workshops. Anna is a Fellow of the Higher Education Academy (UK) and a member of the Management Committee in several COST Actions. She has also been involved in and/or led several nationally and internationally funded research projects. In addition, she is a member of the editorial boards of the *International Journal of Contemporary Hospitality Management, Tourism Management Perspectives, Tourism Review* and the *Journal of Service Theory and Practice*. Her research interests lie primarily in the areas of tourism planning and development with emphasis on sustainable tourism and tourist behaviour.

Dimitirios Stylidis
University of Crete, Greece
Email: d.stylidis@uoc.gr

Dimitrios Stylidis is an Associate Professor in Tourism Marketing at the University of Crete, Greece. Prior to this, he was a Senior Lecturer in Tourism and Research Cluster leader (place image, identity and heritage) at Middlesex University London, and a postdoctoral researcher and Visiting Lecturer at Ben-Gurion University of the Negev, Israel. He holds a PhD and an MSc in Tourism Management from the University of Surrey, UK. He is the Research Notes Editor of the *International Journal of Tourism Policy*. His work has appeared in leading academic journals including the *Annals of Tourism Research, Journal of Travel Research* and *Tourism Management*. His current research interests include place image, tourism impacts and residents' attitudes towards tourism development.

Contributors

Wisam Abbasi is a senior freelance consultant in tourism and sustainable development for international and local clients. She is an accomplished adviser in the field of community-based and alternative tourism. She graduated from the Birzeit University, Palestine, in 2001 with a bachelor's degree in Business Administration and has a Master's degree in Tourism Development from Bethlehem University. She is currently a PhD student at SOAS, University of London, in the department of Social Anthropology. Email: wisamabbasi@outlook.com

Naser Abdi has a BA in Tourism Management and an MA in Tourism Planning. Naser is a PhD student in Tourism at the University of Algarve, Faro, Portugal. He has researched tourism-related topics including cultural tourism and tourism geography. He also is a member of various tourism-related committees in Iran. Email: naserabdi1989@gmail.com

Elad Almog is a Director of Tourism with a demonstrated history of working in the leisure, travel and tourism industry. He is experienced in the hospitality industry, event management and marketing. His PhD, from Ben-Gurion University of the Negev, focused on Geography and Environmental Development. His research focuses on the establishment of tourism routes and trails, regional development and tourism planning. Email: dr.almoge@gmail.com

Selenay Ata has an MA in Sustainable Tourism Management from Boğaziçi University in Istanbul, Turkey. She has taken part in research projects on corporate social responsibility, tourism as a tool for sustainable development and local involvement. Her research interests also include sustainability, heritage management, local resident perceptions and tourism at World Heritage Sites, particularly in rural settings. Email: selencanata@hotmail.com

Nurşah Ayhan obtained her BA in Tourism Administration from Boğaziçi University in Istanbul, Turkey, with a dissertation on place branding and wine culture. She is currently studying for a research degree in Business and Finance at University Carlos 3 in Madrid, Spain. Email: nursahayhan@gmail.com

Fatemeh Bagheri has a BA, MA and PhD in Tourism Management, and is a Professor of Tourism at the University of Guilan, Iran. She has written extensively on tourism marketing, cultural tourism, recovery strategies and entrepreneurship in tourism. As a reviewer, Fatemeh works with several Iranian academic journals publishing tourism research. Email: f.bagheri@guilan.ac.ir

Ashkan Borouj has an MA in Tourism Management. He is very passionate about marketing, innovation and training in the field of tourism. He is a consultant and trainer in strategic marketing and development, and is the founder of the Boomgardi Sustainable Tourism Development Institute. He has over 10 years of experience in training and consulting for small businesses, especially ecolodges and start-ups, for sustainable development. His innovative and creative activities are featured in the United Nations World Tourism Organization book, *Iran Tourism Stories*. Email: ashkanborouj@gmail.com

Ali Hajinejad is an Associate Professor of Geography at the University of Guilan, Iran. He has worked on different projects related to urban planning, and his research focuses on tourism policy making, tourism management process, destination management and strategic approaches to tourism development. Email: ahajinajad@gmail.com

Farid Hanifezadeh is an MA student in Tourism Planning and Development at Allameh Tabataba'i University, Tehran, Iran. His research interests include tourism development, ecotourism, community-based tourism, sustainable development and tourist behaviour analysis. He is currently working on his Master's thesis on the factors that influence visitors' environmentally friendly behaviour. Email: f.hanifezadeh@gmail.com

Rami Isaac has a PhD from the University of Groningen, The Netherlands. He is currently a Senior Lecturer in tourism at Breda University of Applied Sciences in The Netherlands. In addition, he is an Assistant Professor at the Faculty of Tourism and Hotel Management at Bethlehem University, Palestine. His research interests lie in the area of tourism development and management, critical theory and political aspects of tourism. He has published numerous articles on tourism and political (in)stability, occupation, tourism and war, and dark tourism. Email: Isaac.R@buas.nl

Sina Kuzuoglu is a PhD student at the Department of Recreation and Leisure Studies at the University of Waterloo, Ontario, Canada. He completed an MA in Sustainable Tourism Management at Boğaziçi University, Istanbul, Turkey, where he was primarily involved in research projects on tourism in cultural heritage settings, particularly at World Heritage Sites in Turkey. His current research interests are mainly in the cultural offerings of urban areas and their interaction with local communities. Email: skuzuogl@uwaterloo.ca

Sarasadat Makian is a PhD student at the Grenoble Alpes University's Urban Planning and Alpine Geography Institute in France. She graduated from Iran's Allameh Tabataba'i University with a bachelor's and a Master's degree in Tourism Management. She also holds a Master's degree in Tourism, Innovation, and Transition from the University of Grenoble Alpes in France. Her research interests include tourism future studies and destination development. Email: sara.makian@gmail.com

Kholoud Mohamed Abdel Maksoud, PhD is an Associate Professor in the Natural Resources department, Faculty of African Postgraduate Studies, Cairo University, Egypt. Her Research interest include environmental geology, working on Geo-heritage and Geosites assessment, and Geo-archaeology. In addition to this, she has worked on igneous petrology as a branch in her research, and remote sensing applications in different branches of geology. Email: kholoud.mali@staff.cu.edu.eg

Bahram Nekouie Sadry, PhD is a Geotourism Consultant and Adjunct Senior Lecturer of the University of Science and Culture in Tehran, and Chairman of Geotourism Standing Committee, Ecotourism Scientific Association of Iran. He has published textbooks on geotourism in Iran and has undertaken geotourism consultancy projects. Dr Sadry also has designed and conducted geotour-guide training courses in Iran. His latest important academic work is a reference book: *The Geotourism Industry in the 21st Century* (2021) published by Apple Academic Press, Florida. Email: Bahram.Sadry@usc.ac.ir

Alexis Saveriades is an Assistant Professor of Tourism Policy, Planning and Development with the Cyprus University of Technology. He is a University of Surrey graduate with qualifications in the fields of Hospitality and Tourism Planning and Development. Under a Fulbright Commission Scholarship, he also studied in the field of Strategic Management at Cornell University. Alexis enjoys extensive teaching experience at undergraduate and postgraduate level. Furthermore, he is actively engaged in policy formulation and strategic planning at regional and national levels, as well as in the management of tourism destinations, destination governance and metagovernance. Email: alexis.saveriades@cut.ac.cy

Ronen Shay has been involved in tourism and hotel management for over 30 years, worldwide and in Israel, including Hyatt and Hilton in Connecticut, Hyatt New York and in Israel, managed hotels in Eilat such as Hilton Queen of Sheba, and was regional manager of the Fattal hotel chain in Eilat. As part of his duties, he was responsible for hundreds of employees and the entire operation of the hotels he managed. He has passed on his rich experience in the hospitality industry to many students in academic courses. Email: rsay@post.bgu.ac.il; ronen.fm102@gmail.com

Somayeh Zahabnazouri founded the Supporters of the Lut Desert (NGO) in order to further research and investigate the desert's natural heritage in the interest of sustainable tourism. She also set up a website www.savethedeserts.com to promote awareness of geomorphological heritage and conservation activities in Kerman province and the Lut desert in Iran. Recently, she was involved with a group of scientists in a land restoration project to combat climate change in southern Italy. She is also a member of the Geotourism Standing Committee, Ecotourism Scientific Association of Iran. Email: zahabnazouri@gmail.com

Introduction

Stella Kladou[1]* and Konstantinos Andriotis[2]
[1]Mediterranean University, Greece; [2]Middlesex University, London, UK

Tourism development and planning undergo different stages and challenges in different regions (Hovinen, 2002; Andriotis, 2011). Theories and models developed at a universal – yet western – level are usually adapted at a regional, national or local one (Tosun, 1998; Milne and Ateljevic, 2001; Calero and Turner, 2020). Yet, new or emerging topics may be developed locally or regionally as well. Considering the numerous benefits of tourism development (Huayhuaca *et al.*, 2010; Rodiris, 2021), its popularity in some countries is not surprising. In others, however, a more introverted planning may be the priority, often linking such strategies with different political, economic and/or socio-cultural interests (McLeod and Airey, 2008).

Given the historical and cultural richness of countries in the Middle East, as well as the economic development some of them have exhibited in recent years (Nazmfar *et al.*, 2019), tourism planning and development have gained much significance for both scholars and practitioners in the region. Turbulence, conflicts and crises exhibited in the area (Ketter and Avraham, 2010; Mohamed and Alseyoufi, 2018) add further dimensions that need to be incorporated in tourism strategies and planning, and taken into consideration by experts at an institutional, corporate and educational level. Furthermore, in order to effectively deal with aspects of sustainability, visions and strategies in the region need to build upon good practices (Andriotis, 2002; Asker *et al.*, 2010). As a result, a greater understanding is required of the factors influencing decision making on tourism matters as well as of the impacts and implications of sustainable tourism development (Huayhuaca *et al.*, 2010).

This edited book addresses the need to examine tourism development and planning in the Middle East from a sustainability perspective by embracing case studies and examples from the region. The book focuses on some countries in the region which have progressed and experienced a remarkable tourism development over the last two decades, namely Cyprus, Iran, Israel, Jordan, Oman, Palestine, Saudi Arabia, Turkey, Saudi Arabia and Oman. A number of factors made it impossible to publish chapters on other Middle Eastern countries, such as Azerbaijan, Bahrain, Iraq, Kuwait, Lebanon, Qatar, Syria and others. The Introduction and Conclusion attempt to shed light on aspects of tourism planning and development common in the Middle East in general. The book was developed during the pandemic outbreak; therefore, some chapters refer to COVID-19-related challenges. Such reference is not central to most authors' argumentation, given that the inherent challenges of tourism development in the Middle East motivate contributors to explore various aspects of management, development and crisis

*Corresponding author: stellakladou@hmu.gr

in tourism. Through its collection of chapters, the book considers sustainable tourism planning and development from the economic, socio-cultural, environmental and regulatory perspectives. Thus, the book advances understanding of the positive and negative impacts of tourism development as well as how turbulence, crisis, synergies, and the top-down and bottom-up approaches to tourism development are connected to different problems and implications for local communities, the region and the relationship of the Middle East with the rest of the world.

Tourism is one of the largest economic sectors worldwide and, according to the United Nations World Tourism Organization (UNWTO) (2021), in 2019 the Middle East was the fastest-growing region in the world in both arrivals and receipts. However, the COVID-19 (coronavirus) pandemic has resulted in a tremendous decline in tourism movements (Andriotis, 2021; Fontanari and Traskevich, 2021; Cifdaloz *et al.*, 2022; Collins-Kreiner and Shilo, 2022; Dar and Kashyap, 2022; Gupta *et al.*, 2022; Mimaki *et al.*, 2022; Mohammed *et al.*, 2022). The COVID-19 outbreak was an unexpected shock for the global tourism industry and the Middle East region has been greatly affected (Forbes, 2022). For instance, more than 60% of business owners in the Gulf Cooperation Council reported that in 2020 their businesses were impacted by travel difficulties and restrictions (Statista Research Department, 2022). As a result, WTTC (2022) highlights that in 2020, the decrease of the travel and tourism (T&T) contribution to GDP in the Middle East was greater (51.1%), compared to the global average (49.1%).

Table 0.1 illustrates the key tourism statistics of each of the Middle East countries where data are available and reveals a varying contribution of T&T per country. In brief, given the statistics in Table 0.1, it is evident that Middle Eastern countries have undertaken diverse pathways, which has generated mixed results, with some of them having experienced fast development and growth, and others having a lower expansion. For instance, when focusing particularly on the contribution of T&T to GDP in 2019, this ranged from 6% in Iraq to 26.8% in Georgia. The T&T industry in Georgia also leads the way in terms of its contribution to total employment (29.3%), which is almost five times higher compared to the respective contribution in Syria (6.2%). Georgia, however, is the country in the Middle East with the greatest impact of the COVID-19 crisis on employment; one in four jobs in T&T (i.e. 26.5%) was lost during the first year of the pandemic (2019–2020). On the other hand, Qatar appears to be the country that has suffered the least in terms of the contribution of T&T to GDP and international visitor impact. Still, employees in the industry experience a less negative impact of the pandemic in Cyprus: interestingly, the change in the total contribution of T&T to employment due to the pandemic is no more than 1.7%, when, on average, COVID-19 resulted in a 19.3% decrease in the total contribution of T&T to employment in the Middle East. In terms of the impact of international visits, the differences are again extraordinary, with UAE receiving US$45.7 billion in 2019 (as opposed to US$1,014.2 million in Syria), and the visitor spend of total exports reaching up to 66.7% in Lebanon but only 1.5% in Kuwait, where the economy heavily relies on other sectors. Lebanon experienced the worse impact of the pandemic in terms of its effect on the contribution of T&T to GDP (–80.9% in 2019–2020), and Iran in terms of the international visitor impact (–86.7% in 2019–2020). Nevertheless, the numbers in Table 0.1 indicate that tourism has an enormous capacity to provide wealth and employment in the Middle East in general, and it is undeniably an important driver of economic and social development. In order to understand the reasons leading to such figures and differences, analysing both tourism planning and development and resilience to crisis for each country is necessary. This book seeks mainly to address the first point, whereas the Introduction offers some insights in relation to the latter.

Crisis and the impact of crisis in general and on tourism in the Middle East in particular is not something new. The Arab Spring uprisings, starting in the early 2010s, for instance, had a significant impact on economy and tourism (Chapman *et al.*, 2021), and even affected countries that, prior to the uprisings, were considered as stability cornerstones in the region (e.g. Syria). In the 21st century and the 9/11 aftermath, tourists appear reluctant to visit regions that have suffered from terrorism (Lutz

Table 0.1. Key economic tourism and travel (T&T) statistics for Middle East countries (2019–2020).

	Total contribution of T&T to GDP*			Total contribution of T&T to employment*			International visitor impact		
	2019		2019/20	2019		2019/20	2019		2019/20
Country	Receipts Total (USD)	%	% Change	Jobs (000s)	% of total employment	% Change	Receipts USD	Visitor spend of total exports (%)	% Change
Armenia	1,798.2 mn	12.9	–75.1	137.8	13.8	–22	1,484.1 mn	28.4	–82.5
Azerbaijan	3,713.1 mn	7.7	–67.1	395.2	8.0	–21.8	1,974.3 mn	8.3	–82.5
Bahrain	4,830.3 mn	12.8	–54.4	96.6	14.6	–22.1	4,367.2 mn	15.9	–57.8
Cyprus	3,390.2 mn	13.4	–74.2	54.0	13.4	**–1.7**	3,154.5 mn	17.4	–81.5
Georgia	4,573.8 mn	26.8	–72.1	517.2	29.3	–26.5	3,398 mn	38.5	–83.1
Iran	41.8 bn	7.9	–48.6	1,852.0	7.4	–16.6	6.5 bn	3.4	–86.7
Iraq	12,026.3 mn	6.0	–50.2	685.2	7.5	–16.9	3,374.2 mn	4.1	–64.2
Israel	23.6 bn	5.6	–58.6	231.1	5.8	–16.6	30.6 bn	7.2	–73
Jordan	6,909.7 mn	16.3	–72.3	255.6	18.6	–23.4	6,229.1 mn	40.3	–77
Kuwait	7,117.6 mn	5.3	–50.9	133.2	5.8	–21.5	1,118.4 mn	1.5	–49.3
Lebanon	19,041.5 mn	19.4	–80.9	430.4	19.1	–26	15,830.7 mn	66.7	–87
Oman	5,332.6 mn	7	–57.6	150	7	–21.4	3,154.5 mn	6.9	–67.9
Palestine*	n/a	n/a	n/a	n/a	n/a	n/a	n/a	n/a	n/a
Qatar	17,891.5 mn	10.4	–34.5	262.3	12.5	–19.8	13,140.1 mn	14.5	–32.9
Saudi Arabia	79.2 bn	9.8	–38.8	1,584.4	12.2	–10.1	30.4 bn	10.4	–80.4
Syria	1,445.9 mn	8.7	–63.5	163.4	6.2	–22.6	1,014.2 mn	21.2	–76.4

Continued

Table 0.1. Continued

	Total contribution of T&T to GDP*			Total contribution of T&T to employment*			International visitor impact		
	2019		2019/20	2019		2019/20	2019		2019/20
Country	Receipts Total (USD)	%	% Change	Jobs (000s)	% of total employment	% Change	Receipts USD	Visitor spend of total exports (%)	% Change
Turkey	77.6 bn	11.0	–54.2	2,616.9	9.3	–16.3	38.8 bn	16.9	–65.2
United Arab Emirates	305.0 bn	10.1	–60.3	4,268.2	12.0	–23.3	45.7 bn	4.9	–66
Mean	–	11.2	–59.6	–	11.9	–19.3	–	18	**–71.4**
Minimum value	–	6.0	–34.5	–	6.2	–1.7	–	1.5	**–32.9**
Maximum value	–	26.8	–80.9	–	29.3	–26.5	–	66.7	**–86.7**

and Lutz, 2020), thereby highlighting a thin line between specific types of tourism activity (e.g. dark/adventure/volunteer tourism) and travel behaviour. The relationship between country and destination image and the relevant influence on foreign investment and tourism have been explored previously and seem to influence travel behaviour and, in effect, tourism planning and development, among other factors (Campo Martinez and Alvarez, 2010; Stepchenkova and Shichkova, 2017; Lutz and Lutz, 2020).

Having experienced a number of crises over the years does not necessarily imply that local and national tourism authorities in the Middle East have developed a strategy to adequately respond to crisis. The impact of the pandemic on key economic indicators per Middle Eastern country (see Table 0.1) is, in fact, an indicator of how even countries that frequently encounter crisis might not automatically demonstrate capabilities in crisis management. The causes leading to greater impact on some countries compared to others might be attributed to internal factors (e.g. how governments responded to the pandemic and the level of 'forgetfulness' as described in Farmaki, 2021) or external ones. For instance, the coronavirus outbreak increases the health disaster risk (Yang *et al.*, 2020) and, as a result, potential travellers may prefer a destination perceived as safer. Given the relationship between tourism and destination and country image (Nadeau *et al.*, 2008; Kladou *et al.*, 2014), an investigation of different tourism types and their development in the Middle East seems necessary in order to pave the way for resilient, sustainable policies.

Within this context and despite the significance of tourism for the Middle East, there are limited book collections. Instead, there are a lot of published works for individual cases (see, for example, Meydani and Uriely, 2007; Morakabati, 2013; Upadhya, 2016). The most recent book focusing on tourism in the Middle East was edited by Seyfi and Hall (2021). The book focuses on culture and heritage in the Middle East and North Africa, highlighting the importance of such elements for tourism development in the region, and the challenges cultural and heritage sites face. According to the contributions of the book, religion seems to play an important role in connecting Middle East and North African cases and, as such, is the main theme of some chapters, with the remaining chapters focusing on other, mainly tangible, aspects of culture and heritage.

A second book, edited by Timothy (2019), debates and embraces tourism not only in the Middle East but also in Northern Africa. In a similar vein to Seyfi and Hall's (2021) book, this edited book also largely reflects a religious approach. Environmental issues, geography and political forces that have long determined the patterns of tourism demand and supply in the area are other priorities set in this book.

Another book on tourism in the Middle East was edited by Daher (2006). This collection appears somewhat outdated, since following its publication more than 15 years ago, much has changed in the Middle East, not only in terms of the geography but, more importantly, in the conditions affecting tourism, policy, planning and development in the region. This book, in line with the aforementioned two books, also has a primary focus on culture, thus does not tackle all aspects of tourism development, planning and sustainability.

From the above it is evident that published books focusing on tourism in the Middle East offer a mainly cultural and tangible heritage approach, primarily reflecting the role of religion, or are outdated. Our edited book extends beyond the particular issues set by earlier publications by seeking to prioritize sustainable tourism planning, policy and development. Thus, the book will advance the understanding of the positive and negative impacts of tourism development as well as how turbulence, crisis, synergies, and the top-down and bottom-up approaches to tourism development are connected to different problems and implications for local communities, the region and the relationship of the Middle East with the rest of the world.

The book aims to provide a thorough analysis of tourism development and planning in the Middle East and to document case studies and examples highlighting good and bad practices in a wide range of Middle East countries. As such, the book explores various contemporary issues including cultural but also social, environmental, economic and regulatory aspects, as well as their implications for sustainable tourism development. Theories and models of tourism development in the Middle East, rural and urban planning, synergies for tourism development and comparative studies provide thorough insights on tourism development and planning in the region. Thus, the book advances understanding of positive and negative impacts of tourism priorities and development, and relevant implications for local communities.

Outline of Chapters

Despite past studies examining various aspects of tourism in the Middle East, little research has been done in the context of the region. This edited collection aims to examine some case studies providing new insights into tourism development and planning processes. Each chapter is authored by scholars who have done extensive research on tourism development and planning in their respective countries.

This edited volume includes an introduction, seven chapters dealing with tourism issues in one particular country, two chapters analyzing tourism issues in different countries in the region and a concluding chapter. This introductory chapter, written by Kladou and Andriotis, sets the scene and provides a background to the context of tourism in the Middle East. This chapter presents key tourism statistics and figures, thereby profiling the tourism industry in this region, briefly reviews other books focusing on tourism in Middle Eastern countries, and sets out the objectives as well as the contribution of the book. The Introduction ends with a short overview of each chapter, outlining the context, aims, methods and results.

In Chapter 1, Alexis Saveriades deploys a critical approach to examine all plans that have been put into force on the Island of Cyprus since its inauguration in 1960. By identifying tourism planning as an integral part of a destination's development strategy, Saveriades unfolds and critically analyses the tourism plans and tourism success story in an attempt to foresee the future of the destination through its past and endeavours.

Chapter 2, by Elad Almog, examines the challenges of establishing tourism routes (TRs), taking as a case Route 90 South in Israel, which connects Jerusalem and Eilat. The author conducted semi-structured interviews with 15 decision makers from the public sector. His main findings include that the route under study has a significant impact on the ability to link the two cities as a synergistic whole. However, several factors of great variance were perceived by the majority of stakeholders as barriers to cooperation, resulting in each part of the route working independently. Among the main implications of the study is the need to analyse each part of the route as an individual unit before trying to connect them all as a whole. In doing so, three main aspects should be thoroughly investigated: brand status, scale of activity and local standards.

In Chapter 3, Rami K. Isaac and Wisam Abbasi report on the tourism experience of Palestine and especially Bethlehem, focusing on women's empowerment. This exploratory study conducted 30 semi-structured in-depth interviews with tourism entrepreneurs and representatives from the pubic sector (women and men) to examine the factors leading to women's low participation in community-based tourism. At the same time, the study attempts to identify the challenges for women's participation and highlight initiatives that could be taken to enhance it and lead to women's empowerment through tourism. The findings of the study reveals that tourism projects do not take gender integration into serious consideration and that the main factors that have a large impact on the limited participation of Palestinian women in community-based tourism projects are culture and traditions, women's social status and Israeli occupation.

In Chapter 4, Sarasadat Makian, Ashkan Borouj and Farid Hanifezadeh analyse sustainable tourism development in rural areas taking as a case small community-based lodges in Iran. The chapter examines the effectiveness of locally owned lodges in supporting sustainable development of community-based rural tourism by applying Delphi and Analytic Hierarchical Process (AHP) methods. In doing so, four core pillars emerged, namely (i) indigenous tourism products, services and activities; (ii) ecological and environmental structure; (iii) family ownership and management and local community participation; and (iv) tourism infrastructure. At the same time, 23 sustainability indicators for community-based lodge development were identified and contribute to rural and remote areas that seek to develop sustainable community-based rural tourism.

In Chapter 5, Fatemeh Bagheri, Ali Hajinejad and Naser Abdi use a quantitative thematic analysis of Persian-language websites to explore tourism officials' and authorities' views on heritage conservation challenges across Iran. Through a careful review of the textual information of the websites and the use of NVivo 8 software, the authors adopted a thematic analysis and identified in total 376 references, clustered in 18 basic themes and four organizing themes, namely local people, organizational, technical, and legal themes. The authors emphasize aspects categorized under soft and hard challenges, and provide an understanding of heritage conservation challenges for tourism development in Iran. The chapter particularly helps to identify heritage conservation barriers that do not allow specific cities to introduce their local identity to domestic and international tourists, thereby adding value to Iran but also other Middle Eastern or developing destinations prioritizing heritage tourism.

Selenay Ata and Sina Kuzuoglu, in Chapter 6, focus on the governance aspect of tourism and conservation at the mixed World Heritage Site of Cappadocia (Turkey), to explore conservation and residents' subjective well-being from a tourism perspective. This quantitative study modified the Happiness Index proposed by Musikanski *et al.* (2017) to include residents' perceptions on tourism and conservation in Cappadocia. A convenience sampling approach was adopted to select 178 residents. The findings reveal a significant association between residents' and governmental institutions' perceptions on Cappadocia's environmental quality. However, the importance attached to the governmental bodies in the environmental and cultural conservation in Cappadocia indicates a disconnect between residents and decision-making processes concerning conservation in the region.

Along the same lines, Nurşah Ayhan and Stella Kladou describe in Chapter 7 how sustainable tourism development goes hand in hand with place identity, given that strategies build upon the natural, cultural and identity characteristics of the focal place. However, stakeholder groups and their profile transform, and identity is also subject to change. Focusing on the island of Bozcaada (Tenedos) in Turkey and its association with wine, the authors argue that the local identity may differ from the national one, thereby leading to the question of which identity to evolve around. This chapter aims to offer an empirically grounded conceptualization of the dialogical relationship between the identity-based approach to place branding and sustainable development. The study combines secondary information with data collected through interviews and focus groups with local stakeholders to describe the path from early sustainable development of wine tourism and identity-based branding to the prioritization of sun–sea–sand tourism initiatives. Based on the case study, the chapter presents and explores an empirically driven albeit conceptual model, which discusses the dialogical relationship of the identity-based approach to place branding and sustainable tourism development, and points out how institutions emerge and influence this dialogue, and how stakeholders can achieve co-creation.

In Chapter 8, Ronen Shay conducts a cross-cultural study of the global multicultural tourism workforce having as cases Israel (Eilat) and Jordan (Aqaba). Drawing on interviews, focus groups, observations and document analysis, the author reviews the 'Eilat project', which permits Jordanians to cross the border to work in the hospitality industry of Eilat. At the same time, the study examines relevant opportunities and shortcomings, as well as the replicability of the project. To illustrate tourism development through human encounters the analysis yields five themes, namely cultural aspects, Israel and Jordan relations, interpersonal interactions, human resource management and entrepreneurship. Findings suggest that success lies in building a solid relationship through mutual understanding and integrating diverse notions and faiths. The study concludes that realizing the cultural base through close and personal encounters, gratifying gestures and pleasant experiences can overcome apprehensions, concerns, prejudice and fear, and ultimately form warmer relations between states.

Chapter 9, by Bahram Nekouie Sadry, Kholoud Mohamed Abdel Maksoud and Somayeh Zahabnazouri, explores cases of geotourism (abiotic nature-based tourism) in four Middle Eastern countries (Saudi Arabia, Oman, Jordan and Iran). The chapter seeks to review the pertinent literature on geotourism studies in the Middle East by investigating the conceptual and historical backgrounds of geotourism development in the region. The study reports that geotourism in the four countries has witnessed significant development and awareness in the Middle East in the last decade and plays a vital role for the local communities' poverty alleviation in adjacent geosites and geoparks.

The Conclusion, by Anna Farmaki and Dimitrios Stylidis, reviews the previous chapters and outlines the main reasons which made planning and policy essential to tourism development and planning in the Middle East. The main conclusion drawn is that adaptations in planning and development at the business and destination level will be instrumental to sustain growth and support the economic recovery of Middle Eastern countries following the COVID-19 pandemic. Regardless of the type of tourism, specific country and destination image characteristics (e.g. safety and hygiene) emerge as particularly important, thereby calling destination planners and policy makers to act accordingly. The chapter also provides directions for future research.

References

Andriotis, K. (2002) Options in tourism development: conscious versus conventional tourism. *Anatolia* 13(1), 73–85.

Andriotis, K. (2011) A comparative study of visitors to urban, coastal and rural areas. Evidence from the island of Crete. *European Journal of Tourism Research* 4(2), 93–108.

Andriotis, K. (2021) Introduction. In: Andriotis, K. (ed.) *Issues and Cases of Degrowth in Tourism*. CAB International, Wallingford, UK, pp. 1–21.

Asker, S., Boronyak, L., Carrard, N. and Paddon, M. (2010) *Effective Community Based Tourism: a Best Practice Manual.* Asia-Pacific Economic Cooperation, Singapore.

Calero, C. and Turner, L.W. (2020) Regional economic development and tourism: a literature review to highlight future directions for regional tourism research. *Tourism Economics* 26(1), 3–26.

Campo Martínez, S. and Alvarez, M.D. (2010). Country versus destination image in a developing country. *Journal of Travel & Tourism Marketing* 27(7), 748–764.

Chapman, C.M., Miller, D.S. and Salley, G. (2021) Social disruption of the tourism and hospitality industries: implications for post-COVID-19 pandemic recovery. *Worldwide Hospitality and Tourism Trends* 13(3), 312–323.

Cifdaloz, E.D., Semiz, B., Parlak, B.O., Yangu, G. and Acar, S. (2022) Input–output effects of the Turkish hospitality industry in the Covid-19 era. *International Journal of Tourism Policy* (in press).

Collins-Kreiner, N. and Shilo, S.D. (2022) The return of the 'Black Swan'? Christian pilgrimage to the Holy Land and the Covid-19 pandemic. *Qualitative Research in Tourism* (in press).

Daher, R.F. (ed.) (2006). *Tourism in the Middle East: Continuity, Change, and Transformation*. Channel View Publications, Bristol, UK.

Dar, H. and Kashyap, K. (2022) Wellness travel motivations in the wake of COVID-19. *International Journal of Tourism Policy* 12(1), 24, 43.

Farmaki, A. (2021). Memory and forgetfulness in tourism crisis research. *Tourism Management* 83, 1–4.

Fontanari, M. and Traskevich, A. (2021) Consensus and diversity regarding overtourism: The Delphi-study and derived assumptions for the post-COVID-19 time. *International Journal of Tourism Policy* 11(2), 161–187.

Forbes (2022) Covid in the Middle East: UAE fights Omicron with boosters – and bossy robots. Available at: https://www.forbes.com/sites/christopherelliott/2022/01/22/covid-in-the-middle-east-uae-fights-omicron-with-boosters-and-bossy-robots (accessed 23 February 2022).

Gupta, S., Shukla, V. and Pandiya, B. (2022) Tourists motivation towards destination visit intention post-pandemic: scale development and validation. *International Journal of Tourism Policy* 12(2), 160–178.

Hovinen, G.R. (2002) Revisiting the destination lifecycle. *Annals of Tourism Research* 29(1), 209–230.

Huayhuaca, C., Cottrell, S., Raadik, J. and Gradl, S. (2010) Resident perceptions of sustainable tourism development: Frankenwald Nature Park, Germany. *International Journal of Tourism Policy* 3(2), 125–141.

Ketter, E. and Avraham, E. (2010) Online tourism marketing for sub-Saharan African countries: battling stereotypes of high risk, hostility and underdevelopment. *International Journal of Tourism Policy* 3(4), 318–331.

Kladou, S.A., Giannopoulos, A. and Assiouras, I. (2014) Matching tourism type and destination image perceptions in a country context. *Journal of Place Management and Development* 7(2), 141–152.

Lutz, B.J. and Lutz, J.M. (2020). Terrorism and tourism in the Caribbean: a regional analysis. *Behavioral Sciences of Terrorism and Political Aggression* 12(1), 55–71.

McLeod, M.T. and Airey, D. (2008). The politics of tourism development: a case of dual governance in Tobago. *International Journal of Tourism Policy* 1(3), 217–231.

Meydani, A. and Uriely, N. (2007) Policy and law in Israel: the case of the Tourism Services Act 1976–2002. *International Journal of Tourism Policy* 1(3), 203–216.

Milne, S. and Ateljevic, I. (2001) Tourism, economic development and the global–local nexus: theory embracing complexity. *Tourism Geographies* 3(4), 369–393,

Mimaki, C.A., Darma, G.S., Widhiasthini, N.W. and Basmantra, I.N. (2022) Predicting post-Covid-19 tourists' loyalty: will they come back and recommend? *International Journal of Tourism Policy* 12(1), 1 – 23.

Mohamed, T.Z.F. and Alseyoufi, T.S. (2018) Evolution and impact of terrorism in the Middle East: implications for Egyptian travel and tourism. *International Journal of Religious Tourism and Pilgrimage* 6(3), 86–106.

Mohammed, A.H.A., Goodman, W.M. and Yfantidou, I. (2022) The impact of airlines' policies during COVID-19 on traveler's repurchase intention. The case of Aegean Airlines. *International Journal of Tourism Policy* 12(2), 133–159.

Morakabati, Y. (2013) Tourism in the Middle East: conflicts, crises and economic diversification, some critical issues. *International Journal of Tourism Research* 15(4), 375–387.

Musikanski, L., Cloutier, S., Bejarano, E., Briggs, D., Colbert, J., Strasser, G. and Russell, S. (2017) Happiness index methodology. *Journal of Social Change* 9(1), 4–31.

Nadeau, J., Heslop, L., O'Reilly, N. and Luk, P. (2008) Destination in a country image context. *Annals of Tourism Research* 35(1), 84–106.

Nazmfar, A., Eshghei, A., Alavi, S. and Pourmoradian, S. (2019) Analysis of travel and tourism competitiveness index in Middle-East countries. *Asia Pacific Journal of Tourism Research* 24(6), 501–513.

Rodiris, L.J.T. (2021) Tourism as a driver of development: an evidence from selected tourism stakeholders in Vigan City. *Tourism and Sustainable Development Review Journal* 2(1), 39–48.

Seyfi, S. and Hall, C.M. (eds). (2021) *Cultural and Heritage Tourism in the Middle East and North Africa: Complexities, Management and Practices*. Routledge, London.

Statista Research Department (2022) Inbound visitor growth in the Middle East from 2011 to 2023. Available at: https://www.statista.com/statistics/311575/inbound-visitor-growth-the-middle-east/ (accessed 23 February 2022).

Stepchenkova, S. and Shichkova, E. (2017) Country and destination image domains of a place: framework for a quantitative comparison. *Journal of Travel Research* 56(6), 776–792.

Timothy, D.J. (ed.) (2019) *Routledge Handbook on Tourism in the Middle East and North Africa*. London: Routledge.

Tosun, S. (1998) Roots of unsustainable tourism development at the local level: the case of Urgup in Turkey. *Tourism Management* 19(6), 595–610.

UNWTO (2021) *International Tourism Highlights, 2020 World Tourism*. UNWTO, Madric, Spain.

Upadhya, A. (2016) Polity, policy and destination management: an exploratory study of political systems and management of tourism with special reference to the GCC countries. *International Journal of Tourism Policy* 6(3/4), 341–361.

WTTC (2022) *Travel & Tourism Global Economic Impact & Trends 2021.* World Travel and Tourism Council, London, UK.

Yang, Y., Zhang, H. and Chen, X. (2020) Coronavirus pandemic and tourism: dynamic stochastic general equilibrium modeling of infectious disease outbreak. *Annals of Tourism Research* 83, 1–6.

1 Strategic Planning and Policy Making for Tourism Development in Cyprus: An Empirical Case-based Approach

Alexis Saveriades*
Department of Hospitality and Tourism Management, Faculty of Tourism Management, Hospitality and Entrepreneurship, Cyprus University of Technology, Limassol, Cyprus

1.1 Introduction

The chapter stems from a thorough and in-depth exploration of the evolution and growth of the tourism sector on the Island of Cyprus, from the first days of its inauguration in 1960 (when the Island gained its independence from Britain) to its mature standing of today. The chapter navigates the reader through the strategic development plans as these were drawn up by the authorities on the Island covering the 1960–2020 period. In turn and through the deployment of pertinent facts, a critical view is presented.

On the theoretical axis, the main pillars this empirically focused chapter investigates emanate from strategic planning, policy making, tourism development and tourism governance, also taking into consideration the evolution of the sector at large over the past six decades. On the ground, the concise and confined setup of a small Island with commendable growth in tourism over this period is thoroughly scrutinized.

For the reader who is not familiar with the Island of Cyprus, its turbulent history and the evolution of its tourism industry, a brief overview of its history is presented at the opening. The chapter then unfolds chronologically with the presentation of the Island's roadmap in strategic planning and policy making for the development of the tourism sector over the past six decades. Milestones are based on the strategic plans that were drawn up over this period of time. For each stage (from the early years to maturity) the key provisions of over this period of time. Under each stage (from the early years to maturity) the key provisions of the plans are unfolded and analysed by interpreting information based on evidence and experience.

The main reference priorities revolve around the strategic plans that have been drawn up over the past 60 years and the empirical contribution of the author, who brings to the surface many pitfalls that have hindered or delayed the turning of the destination into a sustainable one. The reader stands to benefit from this chapter through the holistic view presented of how a destination has evolved throughout this long period. This analysis of the tourism sector in Cyprus would certainly serve as a good example of a destination that is currently in the euphoric stage of growth.

The chapter concludes by looking into the future through the past. Implications for

*alexis.saveriades@cut.ac.cy

DOI: 10.1079/9781789249149.0001

scholars and industry practitioners are presented, limitations are brought to surface and future directions are suggested.

1.2 Brief Background

Tourism plans, whether these are drawn up to serve at local, regional or national level, offer rich, in-depth and in-breadth material which is of importance to practitioners and scholars alike. Tourism plans at national level are of paramount importance because they serve as a roadmap towards development, indicating to all engaged partners and stakeholders that a clear, well-thought-out vision is in place within predetermined objectives and goals, which is supported by developed action plans.

Tourism development plans are important to the business world and of utmost importance to interested investors. On the one hand, an entrepreneur can follow up on the deployment of a plan so as to build a strategy within the tourism sector, while on the other hand, tourism plans are important to the central government, which has to amalgamate these into its national priorities in an attempt to optimize their potential contribution for the benefit of the economy, the environment and society. Furthermore, the study of plans and their connection to public–private partnerships, responsible and sustainable development, international cooperation and competition is of paramount importance to the governance of the sector.

Tourism development plans are equally important to scholars who are eager to investigate whether plans drawn up on paper are applicable in a real setup under different conditions. Additionally, research and evaluation on strategic plans vis-à-vis their success in implementation contribute to progressively minimizing the negative issues and capitalizing on positive outcomes. On the other hand, the absence of planning or the masking of plans raises concerns and, as such, does not serve public benefit.

Tourism undoubtedly has a profound impact on destinations throughout the world. In the quest of earning more 'benefits' from the influx of tourism, many destinations have opted for an *ad hoc* development of the sector or, in cases where a strategic plan has been deployed, this does not really have a strategic orientation, with an overall guiding framework and predetermined strategies towards development objectives, being absent (Hall, 1998). As a consequence, the phenomena of overcrowding, uncontrolled and unplanned development, mass tourism, degradation of quality of life of the indigenous population, alteration of the physical environment, extensive use of scarce resources, exceeding the destination's carrying capacity, inflated land prices etc. are evident throughout. Hence, failing to plan and control the development of tourism within a sustainable context has brought about far too many negative issues that are difficult, if not impossible, to reverse in the fullness of time. Failure to take the big picture into consideration leads to myopic plans that merely serve the sector and the country's economy in a short-term manner. As supported by Ritchie (1999), tourism planning and development decisions need to adopt longer-term perspectives, as the cumulative effects of today's development decisions will have impacts well beyond the lifetimes of those making the decisions.

For the purpose of this empirical case-study chapter, a small in size, yet popular eastern Mediterranean destination was selected because of its maturity stage in terms of tourism development, its rich access to documents and data pertinent to tourism strategy and development plans, as well as the deep knowledge and engagement of the author in the Island's tourism system. Although islands exhibit weaknesses regarding their development perspectives, experience shows that they succeed from an economic perspective. Is economic prosperity the utmost goal in the development of the sector or is tourism development seen as a vehicle of capitalizing on economic prosperity to help towards the island's holistic development while at the same time reducing the strain inflicted upon the social and environmental fabric?

Towards this end, thorough desk research preceded the writing of this case, which entailed a situation analysis of the economic, environmental and physical aspects of planning on the Island, the structure of the tourism sector with all its key actors and stakeholders, as well as the behind-the-scenes channels influencing and bypassing the planning process to the benefit of the few.

1.3 The Island of Cyprus – A Brief Overview

Cyprus (Fig. 1.1) is the third-largest island in the Mediterranean (after Sicily and Sardinia), located in the extreme northeastern corner of Mediterranean, at the crossroads of Europe, Asia and Africa. The Island lies 800 km east of Greece, 300 km north of Egypt, 105 km west of Syria, and 71 km south of Turkey. It measures 240 km latitudinally and 100 km longitudinally. The country is an independent sovereign republic (since 1960) with a presidential system of governance. Its legal system is based on English common law with civil modifications. Cyprus is a full member of the European Union and part of the Eurozone. The Island occupies an area of 9251 km^2 of which 36.2% has been under military occupation by Turkey since 1974. Its 2018 population in the government-controlled areas was estimated at 847,000 distributed among Greek Cypriots (78.6%) and non-Greek Cypriots (21.4%, which includes foreign citizens residing in Cyprus) (Saveriades, 2014a; Republic of Cyprus, 2019b). The two major nationalities that dominate the Island's foreign national community are the British and the Russians.

The Island has a central plain with mountains to the north and south and scattered plains along the southern coast. Cyprus has 648 km of coastline. Its climate is characterized by long, warm and dry summers and mild winters. The all-year-round average temperature is 23°C.

The official languages are Greek and Turkish with Greek being the main language on the Island. The English language is spoken widely throughout and used in business.

The Island's economy is small, open and robust, with the services sector (tourism, financial services, real estate, shipping) constituting about 87.1% of GDP in 2018, 21.9% of which is attributed to travel and tourism. In 2018, around 10.6% of the employment force (437,500) was directly engaged in the tourism industry. The GDP per capita for 2018 was estimated at €23,846.80 (Republic of Cyprus, 2019a, 2020a; STATISTA, 2018). The literacy rate is at 97.6% while crime is at 0.90 per

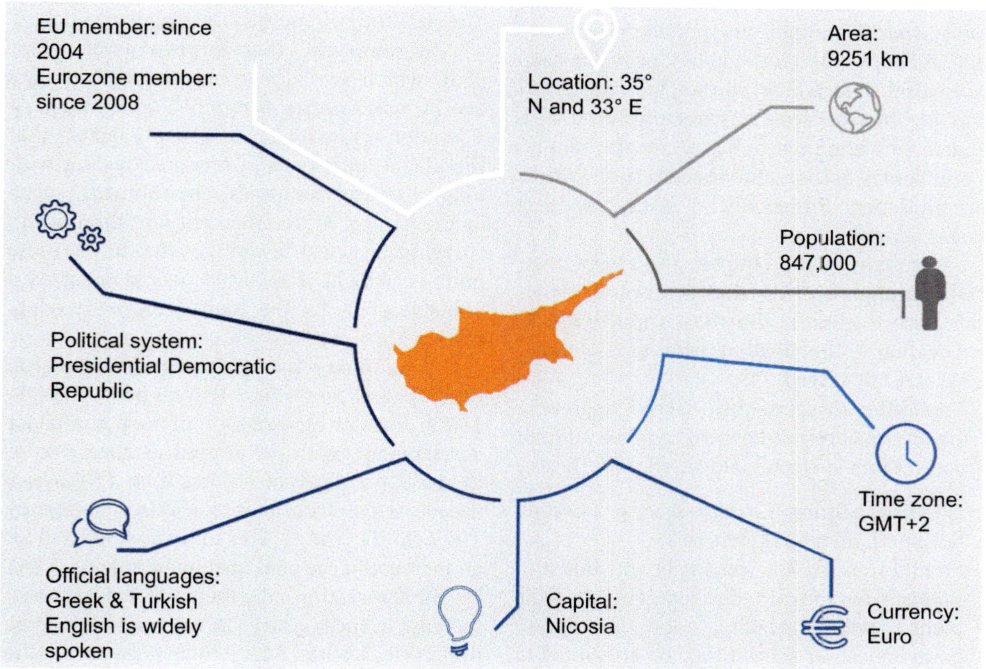

Fig. 1.1. Cyprus: key facts and figures. Source: Cyprus Investment Promotion Agency, 2020.

100 persons. According to the World Health Organization (2017) Cyprus is the safest country in the world for young people out of 184 countries across the globe.

1.4 Methodology

A qualitative empirical case study approach was selected for this field research. A thorough study of the actual strategic plans over a 60-year period, coupled with the author's deep knowledge of and engagement with the Island's tourism industry, is at the forefront of the methodology deployed.

The chapter does not aim to produce statistical associations between variables. As a qualitative empirical study, it aims to provide information that is rich in depth and breadth to aid the reader to understand the happenings within the specific boundaries of the research area (the Island of Cyprus). Findings cannot be generalized or transferred to other locations *per se*, but can serve as highlights of the aftermath of decisions taken at a specific point in time, under specific conditions prevalent at that time. As such, the methodology provides deep contextual data which cannot be transferred as is to other setups but can only serve as a guideline by bringing to the surface the 'don'ts' if one comes across similar scenarios. Within this analytical framework, a critical writing style is adopted.

Throughout the chapter, the deployment of the analysis ensures that descriptive (what, when, who, where), analytical (how, what if) and evaluative (implication, what next) components are duly served.

In order to serve the selected approach, two questions are put forward, and an attempt is made to provide justifiable answers for these.

a. Is Cyprus a tourism success story or has it all been built on false pretences?
b. Should tourism flow results be attributed to noteworthy tourism development strategic plans, policy making and exemplary governance? Or should they be attributed to conjunctures that have worked towards this end?

1.5 Strategic Planning and Policy Making for the Development of the Tourism Sector: Six Decades in Retrospect

1.5.1 Evolution and the early years (1960–1973)

On 16 August 1960, Cyprus gained its independence from the United Kingdom. Until then and since 1925, Cyprus had served the UK as its Crown Colony.

While some tourist activity and subsequent infrastructural development can be traced during Cyprus's rule under Britain, i.e. before 1960, the evolution of Cyprus tourism commenced in 1962 with the drawing up of a purposefully designed and exploited development plan. As an independent nation, Cyprus, and specifically the Ministry of Commerce and Industry (at the time, a tourist office operated under the Ministry), called in a group of expert architects from the Société Centrale pour l'Equipment du Territoire Cooperation, under the auspices of the French Ministry of Foreign Affairs, to formulate the first tourism development plan for the Island, entitled 'Cyprus Study of Tourism Development'.

In retrospect, that tourism development plan, which was drawn up almost 60 years ago by non-tourism personnel, took into consideration many major influencing factors that the global tourism industry is struggling with today (natural resources, environment, social fabric, controlled development and the Island's carrying capacity), bringing to the surface those notions, which, if followed, would serve as a valuable guide for the Island's tourist development within a sustainable context.

In summary, the 'Cyprus Study of Tourism Development', covering a 10-year period (1963–1973), foresaw an explosion in the potential for development, with an annual average rate of increase in arrivals of not less than 15% over a decade, which translates to 150,000 visitors by 1972 (MCIT, 1962). This in turn would call for an increase in the bed capacity by at least 5000. On the forecasting side, the study envisaged an increase in the country's gross national income by 12% in 5 years. At the time of preparing the plan (in 1961), tourist arrivals on the Island were 40,140, with visitors at the time coming mainly from the UK, Israel, Greece, Egypt and the US. Furthermore, the plan called for the setting up of a trust organization (in the form of a joint business government company comprising state, public and semi-public institutions) that, in collaboration with the Tourist office, would secure the financing of various operations and funding for development projects. The plan called for development to adhere to the protection of certain areas. It explicitly states 'There are many beautiful sites in Cyprus; their number is not, however, inexhaustible. As they are expected to be the main power of attraction, they must on no account be spoilt' (MCIT, 1962, p. 8). Towards this end, the plan called for the drawing up of detailed legislation on the preservation of character, the preservation of flora and fauna and the protection of sites through the creation of prohibition regulations.

One has to acknowledge the fact that in those years, tourism planning was not a fully developed scientific field, hence, at the time, town planners and architects were assumed to be the experts, or the closest possible to experts, in the field. One has also to acknowledge that while the field of spatial and town planning and the field of tourism planning are distinct areas of scholarly and scientific interest, they go hand in hand when it comes to sustainability in the development of tourism.

At the heart of this strategy lie a series of regional and local studies, which portray in detail their characteristics and bring to surface what and where specific tourism development projects could help build the Island's tourist product.

Contrary to the fact that infrastructure was, at the time, weak to support middle-to-high-end tourism, on the ground, government policy was geared towards attracting high- and middle-income visitors, who were seen to provide the greatest benefit to the economy. Indirectly, mass tourism was excluded (Andronikou, 1979). This was because tourism was, at the time, seen as a cash injector that would in turn help the country finance the development of social services as well as fund the necessary infrastructure projects – clearly a myopic view in retrospect, especially when one takes into account the efforts exerted today to embrace tourism development within a sustainable context. From an economic point of view the plan certainly succeeded in securing an average of 19% (yearly rise) in tourist arrivals between 1963 and 1973, with Kyrenia and Famagusta (the coastal cities/resorts) accounting for over 70% of all arrivals (Republic of Cyprus, 2019a). However, government actions, contrary to the policy that the same government had drawn up, encouraged the rapid (and to a large extent, uncontrolled) and mass development of the industry.

Despite the government's policy to pursue the development of a product geared towards affluent clientele, development on the ground was far from sustainable, with mass coastal development dominating the scene. This is to a large extent attributed to the incomplete nature of the town and planning laws that existed at the time, which did not contain any provisions for tourism development. The amendment of the town and planning laws came into force much later, in 1990. By that time, the uncontrolled development was so broadly spread that it was extremely difficult if not impossible to streamline the sector. The aftermath of this alone is strongly felt, even today.

In 1969, a semi-governmental organization, the Cyprus Tourism Organisation (CTO), was established and was assigned the responsibility for the development and promotion of Cyprus's tourism.

1.5.2 The hard years 1974–1989

The 'success' (in terms of tourist arrivals and revenue from tourism) of tourism development changed abruptly in the summer of 1974 when Turkey invaded and occupied 37% of the Island's territory. This meant that four-fifths of all accommodation, and virtually all hotels under construction, were removed from the Cypriot economy, as these were located in the northern, now occupied, part of the Island (Andronikou, 1979). Furthermore, Nicosia's International Airport (at the time, the main airport of entry to the Island) ceased operation.

With the economy on the verge of collapse, the government initiated a series of Emergency Economic action plans, which tackled many economic and social parameters, including the re-establishment of tourism in the non-occupied southern part of the Island, as a tourist destination. The plans comprised support schemes mainly under the pillar of economic sustainability. Some environmental (spatial) schemes were also included. Towards this end and in order to maximize tourism development potential, low interest rates, free government land in coastal areas, duty-free imports of hotel material and equipment, the relaxation of planning controls, development of easements, and other economic and fiscal incentives and tax benefits, were introduced to encourage the private sector to take a leading role in rebuilding the Island's tourism (Ioannides, 1991). It was evident throughout that the government was keen to help refugee hoteliers, and to reactivate their businesses on new ground as soon as possible. Concurrently, Larnaca Airport was rehabilitated and expanded so as to re-establish international air transport connectivity.

Although the Turkish invasion had an immediate effect on visitor arrivals, this was comparatively short-lived. Visitor numbers plunged in 1975 to 47,000, but by 1976,these reached 172,000, and by 1979, arrivals surpassed pre-invasion levels (Godfrey, 1996).

A development study (ANFA Consultants, 1976) and three Emergency Tourist Action plans (covering the periods 1976–1978, 1979–1981 and 1982–1986) were drawn up at the time to help the tourist industry recover (Republic of Cyprus, 1976, 1979, 1982). In all, one could read the growth figures and effortlessly conclude that these plans were successful enough to put Cyprus back on the tourism track. However, looking deep down into what has happened, it is evident that this 'success' could be attributed to the targeting of a single market (the UK), which by the end of the aforementioned period accounted for 65+% of tourist arrivals. In order to satisfy the desire for maximum profits, a *laissez-faire* approach prevailed with no notion of regional balanced economic development, no physical and build environment planning and no carrying capacity studies. Capitalizing on the absence of an up-to-date regulatory and planning framework, control of development was virtually lost. In 1980, the government brought forward a new set of regulations in an effort to bring development back under control. These, however, were once again accompanied by another set of incentives. For the CTO, these regulations were (on paper) the foundation for a rationally balanced development. However, the aftermath of the pitfalls described earlier became irreversible and are evident today with a high discomfort level stemming from the uncontrolled development affecting locals and tourists alike.

Contrary to the geographical distribution of tourism development as this was emphasized in the plans, development on the ground was in the coastal areas with Ayia Napa (a traditional fishing village) accounting for 30.7% of the licensed total bed capacity in 1985 (Andronikou, 1987, p. 27). Between 1981 and 1990, bed capacity in the Republic grew from just over 15,000 to nearly 60,000 (Republic of Cyprus, 2019a). This explosion in bed inventory in such a short time was enough to seriously undermine the CTO's efforts to portray the Island as a quality destination (Gillmor, 1989).

By 1990, nine out of every ten licensed tourist beds were located along the Island's southern coast reinforcing the country's dependence on beach-based 'sun-lust' tourism (Ioannides, 1991). Many more unlicensed accommodation establishments escalated the aforementioned problem with social impacts, influx of foreign workers in the hospitality industry and the rise of land prices being marked as the most evident aftermath of this rapid and uncontrolled development. Ayia Napa is a case study of how a traditional fishing village was turned into a disproportionately huge tourist resort where the locals were virtually displaced from their land with total loss of identity and an array of socio-cultural problems (Saveriades, 2007). These problems are to date irreversible.

1.5.3 The golden era 1990–2000

This single activity model revolving around the sun, sand and sea image was, in the late 1980s, raised by the CTO, who questioned its long-term sustainability. The government then began, once again, to rethink tourism on the Island. At that time, the World Tourism Organization (WTO) was called in to help in the preparation of Cyprus's new tourism policy for 1990–2000. The new plan (CTO, 1990) placed a lot of emphasis on building solid and sound foundations at the planning stage, taking into consideration not only economic but also social and environmental factors. The CTO was assigned the responsibility to implement the plan. In 'support' the Government also put into effect the new Town and Country Planning Law (Republic of Cyprus, 1990), which provided the legal framework for all land use. However, in the absence of a strong regulatory framework that would have granted authority to the CTO to enforce and coordinate all prevailing development activities, authority was scattered around tens of different public bodies under different ministries, often with conflicting priorities. As expected, the result was once again detrimental, with the government's inability to regulate the sector a hard and sad fact. Despite the lack of power to control its stakes, the CTO managed, under adverse conditions, to sustain a healthy growth for the sector. If the CTO had been granted the tools to control and fully govern the sector's development, the state of tourism on the Island today would have been more refined, mild, directed and sustainable.

What is questionable even today is whether this can be attributed to the efforts of the CTO or is it because of the image of an unspoilt destination in the sunny Mediterranean that Cyprus enjoyed at the time? It is this author's belief that the CTO played a healthy role in the sustainable development of tourism on the Island at the time. Examples such as the Rural Tourism Initiative launched by the CTO in 1991 and the tourism development of the Akamas Peninsula, which included the 'Laona Project', demonstrate that the CTO intended the sector to develop on solid ground. Furthermore, the CTO drew up other special interest product strategies, such as the golfing strategy and the marine tourism strategy.

The author continues to believe that the rapid and unrestrained development leading up to 1990, coupled with the occasional relaxation of planning restrictions in certain areas by the government, contrary to its own development plan, did not help the efforts to regulate the sector. Certain development projects were given the green light when they should not have been, but we cannot turn back time, and one has to amalgamate all these into the new tomorrow. However, with the altering of the character of the local area, this is easier said than done.

1.5.4 Euphoria (2000–2010)

The guiding vision that inspired the 2000–2010 national tourism strategy is embodied in the following visionary statement: 'Cyprus: a qualitative tourist destination that will satisfy the visitors with various interests and quests and ensure the continuous improvement of the quality of life of the locals' (CTO, 2000, p. 5).

This strategy envisaged that, within the context of sustainable development, actions put forward would safeguard and nurture the quality of the tourism experience, the beauty of the natural environment as well as the quality of life of the indigenous population by making best use of the resources available without depleting, eroding or exhausting them, thus enabling the destination to respond successfully to the needs of visitors and locals alike. Albeit sounding promising enough to secure the long awaited 'right path' for the development of tourism, these bold statements were accompanied by a set of overambitious numerical targets in an urge to increase both the number of visitors and the revenue derived directly and indirectly from their visit and activity on the Island. The plan replaced 'hard' terminology such as 'monetary targets' with soft phrases such as 'socio-economic' benefits.

Nevertheless, the strategy attempted to reposition Cyprus on a pathway that built on the warmth and hospitality of the locals as well as the authentic qualitative experience one would enjoy while visiting the Island. Towards this end, the strategy revolved around three distinct strategies supported by detailed action plans (the marketing strategy, the product strategy, and the quality and value-added strategy). In these, special emphasis is placed on the further penetration into existing markets; the exploitation of new markets; the upgrading of existing products and the development of new ones; the introduction of e-marketing; the development of special interest products to shift away from the traditional sun, sea and sand destination; the creation of a multidimensional experience; the enhancement of guest satisfaction; the cultivation of a distinctive character and identity; the elevation of professionalism levels within the industry; the nurturing of tourism awareness on matters of hospitality and authenticity; and the respect for and protection of the natural and cultural environment and social values. Following the aforementioned, strategies at regional level were conducted following the same tri-pillar approach in line with the national strategy. Towards this end, five regional tourism boards were established in order to implement the regional strategies and their action plans.

The national plan also called for the upgrading and enrichment (in line with the vision of the Strategic Plan for the Development of Tourism) of the necessary infrastructure for which the government and the local authorities are responsible.

Originally the plan spanned 10 years (2000–2010). However, in 2003 a new government was elected to office which wanted to convey a plan of its own. Thus, the existing plan was dropped and rewrapped into a 'new' 7-year plan (CTO, 2003). The differences in the two plans are cosmetic and revolve primarily around a change of the cover and some of the introductory pages. This alone underpins the 'seriousness and professionalism' with which governments and political actors tackle the sector that accounts for one-fifth of the Cypriot economy. It is evident over time that no government since 1960 has credited tourism with the weight it deserves and has elevated it into a national priority, hence, each government draws up plans of its own that seem to bear no relation to the impact of the sector in the fullness of time. It is imperative that the development of the sector should stem from a holistic approach that integrates tourism into the wider context of the economy, society and the environment. For a country in which tourism plays such an important role in its prosperity, tourism should have been elevated into a national priority (similar to education and healthcare). As such, it would follow a predetermined pathway, irrespective of who is in office, which in turn would safeguard long-term planning and sustainability.

The new government at the time assigned the CTO a coordinating role with the mission to ensure that all responsible agents worked towards the implementation of the various tasks under their authority. The hard fact that the CTO had no legal authority over all those (public authorities, local authorities, private sector) who were responsible for their contribution towards the implementation of the strategy caused this otherwise promising strategy to fail.

Despite the fact that, on paper, two bodies were tasked with ensuring the fulfilment of the strategy (one at ministerial level with the five ministers of engaged disciplines being involved in ensuring that their assigned tasks would materialize and the other being a steering committee at CTO level), on the ground, it soon became clear that the ministerial committee was set aside and the strategic plan was left in the hands of the CTO, which lacked the necessary ammunition to fight for its deployment.

One could easily agree with Rhyne (1985), Bryson and Roering (1987), Mintzberg (1994) and Saveriades (2014b) that the 2003–2010 strategic plan for the development of tourism in Cyprus was transformed into a sterile process that bore no relation to strategic thinking and practice. The strategic plan was indeed a figurative and elusive concept.

1.5.5 Maturity 2011–2015

A new 5-year Strategic Plan for Tourism Development (2011–2015) (CTO, 2011) was deployed from January 2011. The major innovative difference in this plan which made it stand out from previous plans, lay in and around the approach adopted. The plan acknowledged the fact that Cyprus was now a mature tourist destination and took into account all the pitfalls of government incompetence in regulating the sector as well as the failure of the preceding plans. As such this 2011–2015 plan revolved around an adaptive model which took into consideration the changes in the global economic scene: the world tourism map and its new and emerging destinations. Stemming from these parameters, it set out to realistically accommodate those projects which would help lift the quality of the services offered and in turn help improve the average spend rather than setting bold numerical targets for tourist arrivals.

Many of the proposed projects that this plan entailed stemmed from the visitor satisfaction surveys that are annually conducted for the CTO. These aim to enhance the value of the tourist experience in an effort to transform the tourist sector into a vibrant, but sustainable one, without necessarily aiming purely at attracting higher numbers of tourists. Furthermore, this strategic plan proposed specific actions in order to help prolong the tourist season in an effort to establish Cyprus as an all-year-round destination with many subsequent positive effects. Special interest products were studied and analysed based on the size of the market, the prospects in attracting targeted groups to engage in these products and the contribution these products would have to alleviating the seasonality factor. Towards this end, a dynamic model was prepared to constantly update the size, flow and impact of these products.

The plan also set qualitative and quantitative parameters in classifying the various target markets (size of market, market penetration, accessibility, product fit, awareness image and perception) and set specific goals for each and every tourist generating market. Moreover, the plan tackled the role of the regional tourism boards as well as the role and development of the tourism industry workforce. Environmental issues within the framework of the model of sustainable tourism development were also addressed. The creations of a 'tourism observatory' was also proposed. Furthermore, this strategic plan unfolds a multi-directional marketing action roadmap with specific marketing and promotional customer-oriented tools geared towards specific needs and wants of targeted groups. This detailed action roadmap is the result of a rigorous evaluation of targeted markets and the behavioural patterns of prospective client groups under different scenarios. Last but equally important, the plan set up a monitoring mechanism to ensure the strategic direction and planned actions were achieved.

Having a subtle and viable plan with all proposed strategies under the control of the CTO led to the promise that with its expertise and will, the organization could bring this strategy to a successful completion.

Two years into the implementation of the plan, the CTO revised some of its actions in line with the prevailing economic situation on the Island as well as to align with prevalent conditions in key source markets as well as in competing destinations. The revised plan was theoretically put into force in January 2013.

At the time, the country was confronted with debt that led to economic and social dislocation. As a consequence of the deteriorating situation for the Cyprus economy, 'Troika' (a term used, especially in the media, to refer to the decision group formed by the European Commission, the European Central Bank and the International Monetary Fund) stepped in to cope with the Cypriot financial crisis. In March 2013 the Economic Adjustment Programme for Cyprus (usually referred to as the bailout programme) was put into force. For the Cyprus Tourism Strategy this meant that all cost-bearing activities would come to a halt and the Troika was necessary in order to proceed. This alone caused a 'forced' retargeting and revision of this sound and visionary strategy to accommodate the demand of Troika for the inclusion of quantitative targets in line with (i) the Island's economic situation, and (ii) the prevailing economic situation in the EU.

1.5.6 Operational plan 2016–2018 (under the control of a memorandum of understanding) (MoU) 2016–2018

In 2016, Troika enforced a 3-year plan for 2016–2018 (CTO, 2016) which was not a strategic plan *per se* but an operational one to ensure (though provisions for 'Good Governance' entailed in the plan) that the tourism sector would abide by the provisions made in the MoU agreed by the government of the Republic and Troika. Along with this, the government of Cyprus decided that the time was ripe to push forward with an amendment of the constitution paving the way for the transformation of the CTO into a Deputy Ministry of Tourism. In order to speed things up and catch up on lost ground, the Government also decided to draw up a new Strategic Plan. Under a tendered competition, the task was assigned to an independent consultancy firm from Spain.

At the same time the Government decided to grant the coordination and project management of this strategy to the office of the 'Presidency of the Republic' and not the CTO which by default is the 'so called' expert authority on tourism on the Island. The 'Presidency' office is staffed by civil servants from the Directorate General for European Programmes (formerly known as the Planning Bureau), an anachronistic government office set up in 1960 to help (at the time) ministries embark on development projects. To date, this office, neglecting altogether the amalgamating, interdisciplinary and complex nature of progressive development at the present time, claims expertise in all possible disciplines. Upon completion, implementation of the strategy was granted to the Ministry of Commerce and Industry, stripping the CTO of its most important role. It is the author's expert opinion that many of the failures in the strategic planning in most economic sectors on the Island of Cyprus can be attributed to the protagonistic role granted *ex officio* to this office. In December 2016, the Spanish consultants delivered a Tourism Strategic Plan to cover the period 2017–2030 (MCIT, 2017). In 2017 this was publicly presented and approved. The Strategic Plan was devised based on the brief that by the time of completion, the Deputy Ministry would be up and running and that this new entity would take over its implementation. However, this was not the case as the Deputy Ministry was not constitutionally formed, hence the newly drawn up strategy was in fact shelved.

The long delay in putting forward the creation of the Deputy Ministry of Tourism had consequentially brought about the freeze of all activity in the development of the sector. The Ministry of Commerce and Industry, totally ignoring the existence and role of the CTO and being incapable of dealing with the strong objection from environmental sustainability groups to the details which surfaced during public consultation on the provision of the plan, took virtually no action on the strategy until the Deputy Ministry of Tourism was created. It was by then evident that 2 years had been lost with no action taken whatsoever on deploying the Strategic Plan.

1.5.7 The new era (2019) and the way into the future 2020–2030

It was not until January 2019 that the Deputy Ministry of Tourism (DMoT) was brought to life. However, by the time the DMoT was formed, the aforementioned strategy completed in December 2016, but not enforced to this day, was already outdated and in need of revision. Unfortunately, and up until the publication of this chapter, no environmental impact study has been conducted, a task which is deemed mandatory under EU law for strategic projects of this nature. As such, the strategy cannot get off the ground, and what is dubbed strategic planning is not a strategy as such.

Contrary to the above mandatory prerequisite, when the Deputy Ministry took over and the CTO ceased operation, the new authority used the 2017–2030 strategy originally drawn up by the Spanish consultants, which had been sitting idle for more than 2 years, to draw up yet another strategy. While this sounds correct in that the strategy was in need of updating, what actually happened was that it was ripped apart, deliberately omitting many of the actions that had been accepted. Hence the new 'strategy' has no cohesion, and most importantly it lacks the development of those concrete actions that could elevate the state of tourism and pave the way towards its sustainable development. As it stands, this document defies all rules and boundaries of what a strategy entails. It is in fact a document with various embellishments, which should have otherwise been dealt with at regional or at local level, that has no specific direction, since it does not stem from a holistic strategy *per se* for the sector. In academia we might classify this so-called strategy as a student group's brainstorming essay that makes for surreal reading. The best-case scenario would be to call this document a local/regional business plan, albeit seriously short-sighted.

Following up on the aforementioned political manoeuvres, today the strategy covers the 2020–2030 period, which, under the auspices, control and coordination of the Deputy Ministry, with all the tools and ammunition needed for its horizontal and vertical deployment, is now in force. Whether the Deputy Ministry succeeds in its new role remains to be seen.

It is beyond the scope of this chapter to comment upon the current plan, which has just got off the ground; however, some alarming points are raised in the text that follows which unfortunately suggest that it is heading in the wrong direction.

The visionary statement, which should otherwise create a very positive climate, sparking change and paving the way forward, is restricted to the cliché: 'Developing Cyprus's tourism in a sustainable way that positively impacts the country, the society and the environment' (DMoT, 2020, p. 21). It goes on to state that the plan envisages a 31% growth in arrivals, to reach 5 million by 2030.

A statement is also made about the evolution of a socially responsible destination; however, there are no provisions for measuring and dealing with the social impact of tourism. Nor is there any visionary direction for the qualitative aspects of the destination, sparingly mentioned in the document. This omission is also evident when looking into the key performance indicators intended to follow up on the deployment of the strategy. All six indicators are quantitative in nature, hence proving, once again, that the aim of this business plan (dubbed 'strategy') revolves solely around arrivals, bed nights, revenue, average spend, seasonal distribution and the destination competitive set, ignoring altogether any qualitative indicators.

Based on information that comes out in a scattered and probably deliberately confusing pattern primarily through press releases and press interviews, there seems to be no coherent and consistent strategic plan in place. Since the strategy is, for some unknown reason, classified as a confidential document, the judgement presented in the text that follows stems primarily from information found in the media. The horizontal role of the DMoT, which should have been at the forefront of the evolution of the CTO into a ministry, is nowhere to be seen. It is this author's belief that with no carrying capacity study, no indices, and no solid and rich time-series data on the Cyprus market vis-à-vis competing destinations and markets to support the proposed actions, there can be no sound strategic visualization and no real differential product strategy. The fact alone that the DMoT aspires to almost doubling the number of tourist arrivals by 2030 at a time when the physical, social and environmental carrying capacity thresholds have never been established, is alarming.

One would have also expected that the promotion of the CTO to a ministry would have brought about internal changes that would render its operation proactive. What has actually happened is that the operation is now fully bureaucratic with little authority to take decisions on the spot, especially when these entail the use of public funds.

What is profoundly evident is a product characterized as a bit of anything and everything, with no differentiated branding product-mix and strategy, with an outdated and to a large extent generic promotional strategy, with weak digital exposure on the international scene, all under a Ministry manned at the top of the hierarchy by people with no scientific exposure to tourism planning, all of which is alarming to say the least.

One only has to note, for example, that for the 2021 budget for the promotion of the Island's tourism, the Deputy Ministry considers the cooperation under joint promotional campaigns with tour operators a panacea, offering direct opportunities for the further 'development and enhancement' of tourism to the Island. Unfortunately, the Deputy Ministry seems unable to realize that this short-sighted remedy lures the destination into the wrong direction and is in opposition to the long-term sustainable development of the sector. It is imperative that those who politically govern the sector realize that the overall planning is obviously heading in the wrong direction with this erratic 'strategy', which, on the one hand, claims to be working on the enrichment and uplifting of the product to attract high spenders, while, on the other hand, is penetrating deeper into the tour operator mass tourism market.

1.6 Implications, Conclusion and Recommendations

What can clearly be seen from Cyprus's 60-year pathway into tourism is that the sector has been, to a large extent, developed on false pretences. The overarching goals of the plans, especially those deployed after the Turkish invasion, revolved around extraordinary numerical aspirations which were never supported by capacity studies. What became progressively evident is that the use of the term 'sustainable' was, and is still being used to mask high and irregular volumes of mass tourist arrivals, high revenues and accelerated returns on investment, fulfilling the expectations of the politically influential groups of tourism business and land/property development stakeholders. Performance indicators of the last few years verify that while visitor numbers are up, daily per head expenditure is down. One only has to compare the 2019 average per person expenditure (€675) to the 2012 figure (€866 in nominal terms and €839 when deflated in real terms) (Republic of Cyprus, 2020b) to realize that performance results are damaging to the future of the sector. For an island of this size, heading into the wrong direction is detrimental to say the least and extremely difficult to reverse.

One has to look into the future through the lens of the past. Experiences to date emanating from the largely unsuccessful implementation of strategic plans of the past (not solely attributed to the CTO but to economic forces that treated tourism as a cash cow in a passionate desire for an almost immediate return on investment) should alarm those appointed to lead the sector. It is obvious that behind the Ministry's pompous approach with the deployment of so-called social media influencers, the use of catchy terminology and the frequent public statements of no real substance lies a lack of scientific knowledge on how a tourism destination is planned and governed. Deeper, though, is the government's urge to satisfy needs and 'reward' their highly influential political followers. Avidly supporting the notion that tourism is a business at all costs is certainly not the way forward. The Ministry should collect, process, analyse and interpret information on national and international forces and proactively respond, adapt, reframe, drive and navigate the market into what is a fiercely competitive environment. As it stands today, the Ministry's approach is myopic to say the least, as it opts to tackle micro-management issues which have piled up over the years in an attempt to hide the evident lack of visionary strategic thinking. Serious tourism planning professionals, not self-acclaimed 'tourism managers', are needed for a sound restart.

Blame for the situation in hand cannot only be attributed to the Deputy Ministry. If things are to change for a better and more sustainable future, the newly established Ministry should also seek stronger support from the government, which, in the last decade, has not taken the impact of the sector seriously enough. The government urgently needs to rethink and recalculate its contribution to the sector. The disproportionately low reinvestment in supporting this sector, which represents one-fifth of Cyprus's economy, is evident (2006 budget in nominal terms: €55.4m, while 15 years later, the 2021 budget was restricted to €45m). Worth noting here is that the budget also bears the Ministry's annual operational expenses. One has to seriously invest in the sector in order to harvest the massive benefits this investment is capable of yielding.

Global competition, innovative strategies in marketing and advertising, the notion of 'value for money', value over volume, customer-driven products and services, value-based pricing; ordinary vs extraordinary, offering experiential hospitality and enhancing the value and other sustainability issues that the destination faces, promotion of true sustainable development, the role of tourism in the quality of life of engaged actors (employees, indigenous populations and tourists), the evolution of technology, the prevailing trend towards staycation, changes in consumer behaviour as a result of the aftermath of COVID-19, destination recovery strategies, crisis contingency plans and the notion of the sharing economy are just a few of the themes that should proactively alarm the leaders of the sector.

The fact that all the information over a 60-year period is publicly available and can be scrutinized in the quest for a better future is commendable, irrespective of the outcome of these plans. What is worrying is the sudden change in this consistent pattern, concealing core information of the current plan, which not only leaves a lot of grey areas and inhibits accountability, but weakens the country's ability to attract serious investments into the sector.

Furthermore, the DMoT has to work hard towards the creation of a brand identity for the destination, based on a well-developed, selective and strong product mix, which will in turn help lay solid and sound foundations to pave the way into the future. *Mellisonant* announcements on the development of new products (slow tourism, climate-friendly holidays, boutique conferences, holistic wellness, micro experiences, immersive travel, etc.), which surface in the media and which stem from catchy buzzwords in a here-there-everywhere approach, should be dropped. Oversegmentation in such a small country should be avoided at all costs. A handful of special products should be prudently selected and developed to uplift the competitiveness of the destination and differentiate the product offering to specific market segments and clients.

Some regions of the destination seem to be suffering the strain of tourism. Transformative strategic tools should be designed to tackle overtourism towards a more sustainable future in an attempt to mitigate overtourism and to promote true sustainable development within the carrying capacity thresholds of the destination.

The depletion of the sector could still be reversed. This can only be achieved if a drastic re-directional model is adopted, away from the traditional sun-and-sea path, away from mass tourism and tour operator overdependence, away from the offering of poor value for money and away from the seasonal nature of the industry, fostering a more diversified, sustainable and higher-quality product for the benefit of the host society and the Island's economy at large. A customer-driven philosophy/approach where the specific needs and wants of the client would become the focus of development of the sector is the only way forward.

The tourism sector cannot sustainably develop with opportunistic and rapid profit-making approaches. The sector will only evolve with a calculated and well-thought-out strategy, empirically and scientifically designed and supported by real tourism experts, incorporating the authentic traits that differentiate Cyprus from other Mediterranean destinations and which revolve around the offering of unique hospitality experiences. The human capital, which is nowhere to be found in the current strategy, should be the foremost asset of any strategic planner in hospitality and tourism.

The COVID-19 pandemic, which brought the tourism sector to an abrupt halt, gives the opportunity to rethink and redesign. As a mature destination, Cyprus should overcome saturation by deciding on which products it should focus on. It is impossible to sustain the sector and develop with a mediocre approach, spending resources on piecemeal development. Furthermore, it is better late than never to conduct a carrying capacity study which would become the benchmark for all future developments in the sector. The strategy should stem from and be fully supported by this carrying capacity study.

References

Andronikou, A. (1987) *Development of Tourism in Cyprus – Harmonisation of Tourism with the Environment*. Cosmos Publishers, Nicosia, Cyprus.

ANFA Consultants (1976) *The Cyprus Tourism Development Study, 1976–1977*. Study commissioned by the Cyprus Tourism Organisation, Nicosia, Cyprus.

Andronikou, A. (1979) Tourism in Cyprus. In: De Kadt, E. (ed.) *Tourism: Passport to Development*. Open University Press, New York, pp. 237–264.

Bryson, J.M. and Roering, W.D. (1987) Applying private-sector strategic planning in the public sector. *Journal of the American Planning Association* 53(1), 9–22. DOI: 10.1080/01944368708976631.

CTO (1990) *Cyprus's New Tourism Policy 1990–2000*. Cyprus Tourism Organisation, Nicosia, Cyprus.

CTO (2000) *Strategic Plan for Tourism 2000–2010*. Cyprus Tourism Organisation, Nicosia, Cyprus.

CTO (2003) *Strategic Plan for Tourism Development 2003–2010*. Cyprus Tourism Organisation, Nicosia, Cyprus.

CTO (2011) *Strategic Plan for Tourism Development 2011–2015*. Cyprus Tourism Organisation, Nicosia, Cyprus.

CTO (2016) *Tourism Operational Plan 2016–2018*. Cyprus Tourism Organisation, Nicosia, Cyprus.

Cyprus Investment Promotion Agency (2020) *Cyprus: Key Facts and Figures*. Cyprus Investment Promotion Agency, Nicosia, Cyprus.

DMoT (2020) *Cyprus's National Strategy for Tourism 2020–2030*. Nicosia, Cyprus.

Gillmor, D. (1989) Recent tourism development in cyprus. *Geography* 74(2), 262–265.

Godfrey, K.B. (1996) The evolution of tourism planning in Cyprus: Will recent changes prove sustainable? *The Cyprus Review* 8(1), 111–133.

Hall, C.M. (1998) *Tourism Development, Dimensions and Issues*, 3rd edn. Addison Wesley Longman, Melbourne, Australia.

Ioannides, D. (1991) The growth of the Cypriot tourist industry: A challenge to planners. In: *Proceedings of the AESOP-ACSP International Congress*, Oxford, UK,

MCIT (1962) *Cyprus Study of Tourism Development (Prepared in French by E. Beaudoin, M. Baud-Bovy and A. Trazos for Société Centrale Pour l'equipement Du Territoire – Cooperation, Translated into English by Nicole and William Beniston).* Ministry of Commerce, Industry and Tourism, Nicosia, Cyprus.

MCIT (2017) *Cyprus Tourism Strategy 2030*. Ministry of Commerce, Industry and Tourism, Nicosia, Cyprus.

Mintzberg, H. (1994) *The Rise and Fall of Strategic Planning*. Free Press, New York.

Republic of Cyprus (1976) *Emergency Tourist Action Plan 1976–1978*. Planning Bureau, Planning Commission, Nicosia, Cyprus.

Republic of Cyprus (1979) *Emergency Tourist Action Plan 1979–1981*. Planning Bureau, Planning Commission, Nicosia, Cyprus.

Republic of Cyprus (1982) *Emergency Tourist Action Plan 1982–1986*. Planning Bureau, Planning Commission, Nicosia, Cyprus.

Republic of Cyprus (1990) *The Town and Country Planning Law (a revision of the 1972 law).*

Republic of Cyprus (2019a) *1960–2019 Annual Statistics on Tourism Arrivals and on Revenue from Tourism*. Statistical Service of Cyprus, Government Publications, Nicosia, Cyprus.

Republic of Cyprus (2019b) Cyprus in figures (2019 edition). In: *General Statistics, Series II, Report No. 24*. Press Information Office (321/2019).

Republic of Cyprus (2020a) *National Accounts 2019*. Ministry of Finance, Government Publications, Nicosia, Cyprus.

Republic of Cyprus (2020b) Statistical Abstracts 2018. In: *General Statistics, Series I, Report No. 64*. Press Information Office (60/2020).

Rhyne, L.C. (1985) The relationship of information usage characteristics to planning system sophistication: An empirical examination. *Strategic Management Journal* 6(4), 319–337. DOI: 10.1002/smj.4250060403.

Ritchie, J.R.B. (1999) Crafting a value-driven vision for a national tourism treasure. *Tourism Management* 20(3), 273–282. DOI: 10.1016/S0261-5177(98)00123-X.

Saveriades, A. (2007) A sociometric examination of perceptions of the impact of contemporary tourism upon society in the tourist resort of 'Ayia Napa', Cyprus. A qualitative approach. *Loisir et Société/ Society and Leisure (Special Edition)* 30(1), 177–195. DOI: 10.1080/07053436.2007.10707745.

Saveriades, A. (2014a) Cyprus, tourism. In: Jafari, J. and Xiao, H. (eds) *Encyclopedia of Tourism*. Springer, Cham, Switzerland. DOI: 10.1007/978-3-319-01669-6_251-1.

Saveriades, A. (2014b) Planning and organization of the tourism sector in Cyprus. In: Costa, C., Panyik, E. and Buhalis, D. (eds) *European Tourism Planning and Organization Systems, Vol II*. Channel View Publications, Bristol, UK, pp. 279–294.

STATISTA (2018) Number of employees in selected tourism industries in Cyprus from 2014 to 2017 (based on data collected from the UNWTO). Travel tourism & hospitality statistics and market data on travel, tourism & hospitality. Available at: https://www.statista.com/map/europe/cyprus/travel-tourism-hospitality (accessed 21 October 2022).

World Health Organization (2017) World health statistics, 2017: Sustainable Development Goals. Available at: https://apps.who.int/iris/bitstream/handle/10665/255336/9789241565486-eng.pdf (accessed 28 August 2022).

2 Challenges of Establishing Tourism Routes: Top Management Perspectives on Route 90 South, Israel

Elad Almog*
Dead Sea and Arava Science Centre, Masada, Israel

2.1 Introduction

This research deals with the decision makers' perspective on challenges to the establishment of tourism routes. A tourism route is a marketing concept and a way for tourists to plan their itinerary (Hardy, 2003). Such routes enable collaborative marketing to be implemented and shared among various stakeholders from several neighbouring regions. Since the late 1980s, tourism routes have become a worldwide phenomenon: their adoption supports economic growth, cultural enhancement, and a more sustainable method of travel that allows for distant regions to connect and tourists to access lesser-known destinations (MacLeod, 2016). Organizations worldwide, both in the developing and developed world, began to view tourism routes as a practical tool for regional development and shared identity which links heritages and economies (Timothy and Boyd, 2015). Two prominent examples are the European Institute for Cultural Routes and Open Africa. The first initiated more than 40 different cultural routes over the past 30 years (seven of them during the 2020/2021 COVID-19 pandemic), from the famous Santiago de Compostela Pilgrim Routes to the Jewish heritage routes and more (Čeginskas *et al.*, 2021). The second, Open Africa, a far-reaching project, connected six countries with almost 60 different routes (Viljoen *et al.*, 2010). As the tourism route concept continues to evolve, many types of routes continue to appear and range from hiking to cycling to driving (Carson and Cartan, 2011; Rogers and Leung, 2020; Le Gouais *et al.*, 2021).

This research focuses on drive routes, which are defined as paved routes (highways, sideways, scenic routes) that typically connect villages, towns, cities, and so on. These routes function as a bridge between different destination brands, acting as an umbrella brand for the supply side and as a route-based journey for the demand side (Hardy, 2003; Goussous and Haddad, 2014; Berti, 2015; Božić and Tomić, 2016; Graf and Popesku, 2016). For the establishment of the route, stakeholders may include direct and indirect tourism businesses, decision makers from the public sector, and NGOs (Carden, 2006; Lourens, 2007). While collaboration with the business sector is important, the literature demonstrates that tourism routes, being long-term initiatives, are largely established by public sector entities and NGOs (Hardy, 2003; Lourens, 2007; Rogerson, 2007; Iles and Prideaux, 2011; Smith, 2011; Myles, 2014; Bregoli *et al.*, 2016). So far, the pre-establishment phase of tourism routes

*dr.almoge@gmail.com

DOI: 10.1079/9781789249149.0002

– which includes the perceptions of relevant stakeholders – remains overlooked.

The aim of this research is to explore prospective challenges to the establishment of a tourism route in order to improve the odds of a successful initiation process. The objectives of this research are as follows: to ascertain how senior stakeholders responsible for the establishment of a tourism route project evaluate and perceive the challenges; to understand how the diversity of the route itself impacts the ability to cooperate; and to identify the factors that might improve the odds of a successful initiation process. In order to tackle this issue, the author conducted a case study of Route 90 South, Israel, which is the subject of a potential tourism route. This route connects two of Israel's leading tourist destinations – Jerusalem and Eilat – with a linear scenic route that passes through two attractive yet different rural regions – the Dead Sea and Arava. Throughout the years, several attempts have been made to drive cooperation between and within these regions.

A new international airport recently opened at the southern end of this particular route, rendering it a suitable case study for a potential tourism route as the region is becoming significantly more connected and attractive. This airport made it possible to land in the south next to Eilat and proceed upwards to Jerusalem, or to head down from Jerusalem and conclude next to Eilat with the ability to fly back home on various international low-cost flights. When factoring such convenience in with the diverse attractions along the route, its varied landscapes and the attempts made for cooperation between the local authorities along the route, a suitable case study for a potential tourism route project materializes. The sections that follow offer a literature review on this subject, followed by methodology, findings, discussion and conclusions.

2.2 Literature Review: Considering Tourism Routes

Defining a tourism route may be a challenging task in and of itself. In their book concerning routes and trails management, Timothy and Boyd (2015) emerged as the first scholars to fully comprehend and distinguish between various types of tourism routes, which range from hiking and cycling routes to motorcycle routes, 4×4 routes and drive routes. There are two main approaches to creating a tourism route: artificially or organically. Artificial routes are built upon stories, cultures, landscapes, fantasy and more. For example, the Brothers Grimm Fairytale Route in Germany connects several of the villages that served as inspiration for the Brothers Grimm stories (Hemme, 2005). This road stretches along some 600 km and, in employing the tourism route concept, becomes a journey for travellers. Organic routes, on the other hand, usually take the form of an historical route that serves various purposes such as trade (e.g. Inca Trail), pilgrimage (e.g. Santiago de Compostela), heritage (e.g. Route 66) and more (Murray and Graham, 1997; Carden, 2006; Zoomers, 2008; Lois-González and Santos, 2015).

Whether artificial or organic, tourism routes may stretch hundreds or even thousands of kilometres, linking regions, countries and at times continents (e.g. Silk Route). They serve as a tool for cooperation, shared identity and enhanced economic benefit that result in a greater market pool (Moulin and Boniface, 2001; Olsen, 2003; Werner, 2003; Shunnaq *et al.*, 2008; Prideaux and Coghlan, 2011). Considering the COVID-19 pandemic, López and Lois-González (2020) emphasize the significance of tourism routes as an 'alternative form of sustainable traveling, far from crowded centres and in full contact with the rural, contributing to the improvement of local economies in a positive way' (p. 549). The research at hand deals with a planned artificial route that connects two significant staging areas in Israel, Jerusalem and Eilat, and passes through two attractive rural regions, the Dead Sea and Arava, with various villages, attractions and scenic landscapes along the way.

2.2.1 Tourism routes and their formation

According to Freeman's (2010) stakeholder theory, a stakeholder is defined as any group or individual that influences and/or is influenced by an organization's goals and achievements. As Freeman developed his stakeholder theory for contexts related to the business world, it is readily adaptable for tourism research and practice. Using this framework, one might demonstrate how whole tourism regions work collaboratively with various public/private groups and individuals in order to manage and market their collective tourism brand and offerings (Byrd, 2007).

In the case of tourism routes and trails, two broad groups of stakeholders affect and are affected by their establishment. The first group is direct stakeholders, who are responsible for the establishment of tourism routes and often belong to senior management in the public sector and/or NGOs (Timothy and Boyd, 2015). Even when the private sector is responsible for the concept and initiation (e.g. Whiskey Trail in Ireland), the public sector steps in for the purpose of enhanced financing and to ensure a balance of interests (Stoffelen and Vanneste, 2016). Moreover, when collaboration involves a geographically large-scale operation, governmental and/or NGO leadership becomes imperative (Sivijs, 2003; Timothy and Boyd, 2015). Such support is particularly crucial to merging and/or reconciling regional interests into inter-regional ones while making certain that no imbalance of power is taking place (Meyer, 2004; McCarron *et al.*, 2015).

The second group of stakeholders that affect tourism development (e.g. consumption and engagement with tourists; Marschall, 2012) or are affected by it (e.g. in quality of life) includes local communities and tourists who travel the route (Hayes and Macleod, 2008; McCarron *et al.*, 2015). Since this research focuses on the establishment of tourism routes, it deals principally with the first group of stakeholders, particularly those senior stakeholders from the public sector and NGOs who are typically at the helm of tourism route initiatives (Timothy and Boyd, 2015).

Tourism routes and their formation include collaborative work of the stakeholders under one agreed and shared tourism route brand that encompasses all other existing brands within the boundaries of the route itself (Olsen, 2003). The brand represents the story of the route, organically (e.g. the rainforest way; Smith, 2011) or artificially (e.g. freedom trail; King and Davis Gatchet, 2018), and is expressed through digital and printed marketing materials, embraced and presented by the stakeholders (Hardy, 2003; Dunne *et al.*, 2021).

2.2.2 Challenges in tourism route stakeholder collaboration

There is a dearth of knowledge on the challenges to establishing new tourism routes, especially in reference to stakeholder perceptions in the pre-establishment phase. The literature reveals that the main challenge to successfully establishing a tourism route initiative is the stakeholders' ability to cooperate out of understanding and agreement for mutual benefit (Hardy, 2003; Lourens, 2007; Rogerson, 2007; Iles and Prideaux, 2011; Smith, 2011; Myles, 2014 ; Bregoli *et al.*, 2016). Several case studies demonstrate that stakeholders tend to doubt the benefits of a tourism route and state that the advantages are not necessarily clear to them (Briedenhann and Wickens, 2004; Smith, 2011; McCarron *et al.*, 2015). While mutually beneficial cooperation leads to the formalization of various institutions ranging from associations (hotels, restaurants, etc.) to DMOs (Destination Marketing/Management Organizations), it is important to remember that conflict is likely to occur and, as Wang and Krakover (2008) describe it, cooperation alongside competition (coopetition). In other words, while stakeholders may understand that cooperation is required for development, they are cautious not to reveal sensitive business information and tactics as they compete with each other. In a successful tourism route, stakeholders understand that collaborative marketing is necessary to the viability of the tourism route as a whole (Damayanti *et al.*, 2017). Challenges to creating this kind of understanding are predominantly described in the literature as human concerns that involve a lack of trust among business partners and a lack of leadership (Smith, 2011). These concerns may lead to low levels of participation in a tourism route and thus a failed initiative (Carson and Cartan, 2011; Myles, 2014). Most notably, the literature shows how a multiplicity of stakeholders increases the complexity of cooperation (de Araujo and Bramwell, 2002; Beritelli, 2011).

A significant question which arises is the effect a diversified tourism route has – or does not have – on the ability to cooperate. Longer routes naturally subsume a greater number of localities and are thus more likely to embody different services and scales of activity. In attempting to connect all the various dots of such a route, the preeminent question emerges: how might a tourism route become formalized, and what might challenge its establishment? Insufficient evidence in the literature makes this difficult to ascertain. While Darabseh (2010) and Smith (2011) suggest that stakeholders from different locales may find it hard to agree and cooperate, the most important factors influencing a route's establishment still remain unclear. In order to explore this matter further, this research examines a potential tourism route in its pre-establishment phase.

2.3 Case Study

Route 90 South (see Fig. 2.1) is a linear 300 km route in the southeastern part of Israel that stretches along the Dead Sea and Arava peripheral regions, which together constitute about 20% of the country's landmass (Garfunkel *et al.*, 1981). These two regions share similar geographical and sociological facets of desert and agricultural community life but differ in their topographies and tourism offerings. Their linear spatial organization, comparable landscapes, and distinctive-yet-complementary tourism and hospitality offerings render them potential subjects for the development of a new tourism route (Krakover, 2014).

The four regional councils of both regions (two in each region) recognize the potential of a shared tourism route. In 2010, they launched a concept called 'Arts and Crafts Along Route 90 South', which culminated only as a web/printed brochure with little to no cooperation between businesses along the route. Cooperation was therefore limited to brochure distribution without any of the specific components required for developing a new tourism route (signage, standardization, etc). In 2015, another tourism route project was advanced solely for the Arava region in an attempt to connect its two regional councils by way of umbrella branding ('Arava Way'). This initiative remained conceptual and resulted in a tourism route for only one of the regional councils (Eilot), thus splitting the Arava in two according to administrative boundaries (updated October 2021).

These attempts, although rudimentary, indicate the potential a tourism route holds for the Dead Sea and Arava regions and are therefore the subject of this research. This study examines a wide framework of government offices, NGOs and regional councils in order to understand the complexity of implementing a tourism route and the initial perceptions that shape its success or failure.

2.4 Research Tools and Sample

In order to explore the perceived challenges of establishing a tourism route in the Dead Sea and Arava regions, the author conducted semi-structured interviews with decision makers from the four regional councils along the route as well as representatives from six leading government offices and five NGOs. The 15 interviews lasted an average of 1 h and were conducted during the summer of 2016. Apart from one official who appointed a lower-level representative to be interviewed in his place, all interviewees gave consent to be interviewed. The author used purposive sampling, selecting specific stakeholders that are officially responsible for tourism and the rural development of these regions. Each interview took place in the interviewees' offices or at any other location and time subject to the availability of each interviewee. As the stakeholders are all official representatives of the public sector and NGOs, the interviews were also official and lacked the need for anonymity. That being said, the representatives' names are not shown in this research, only their role.

The interview agenda comprised three sections. The first included questions related to experience with cooperation between the regional councils along the route. The second dealt with current situations and plans for cooperation. The third section focused on the possibility of establishing a tourism route for the Dead Sea and Arava regions. This allowed for a rich understanding of the collaborative dynamics experienced by those regions' decision makers while also configuring how these dynamics shape stakeholders' perceptions of potential collaboration on a tourism route project. This process enabled the expression of a wide range of views. Since the investigation involved three

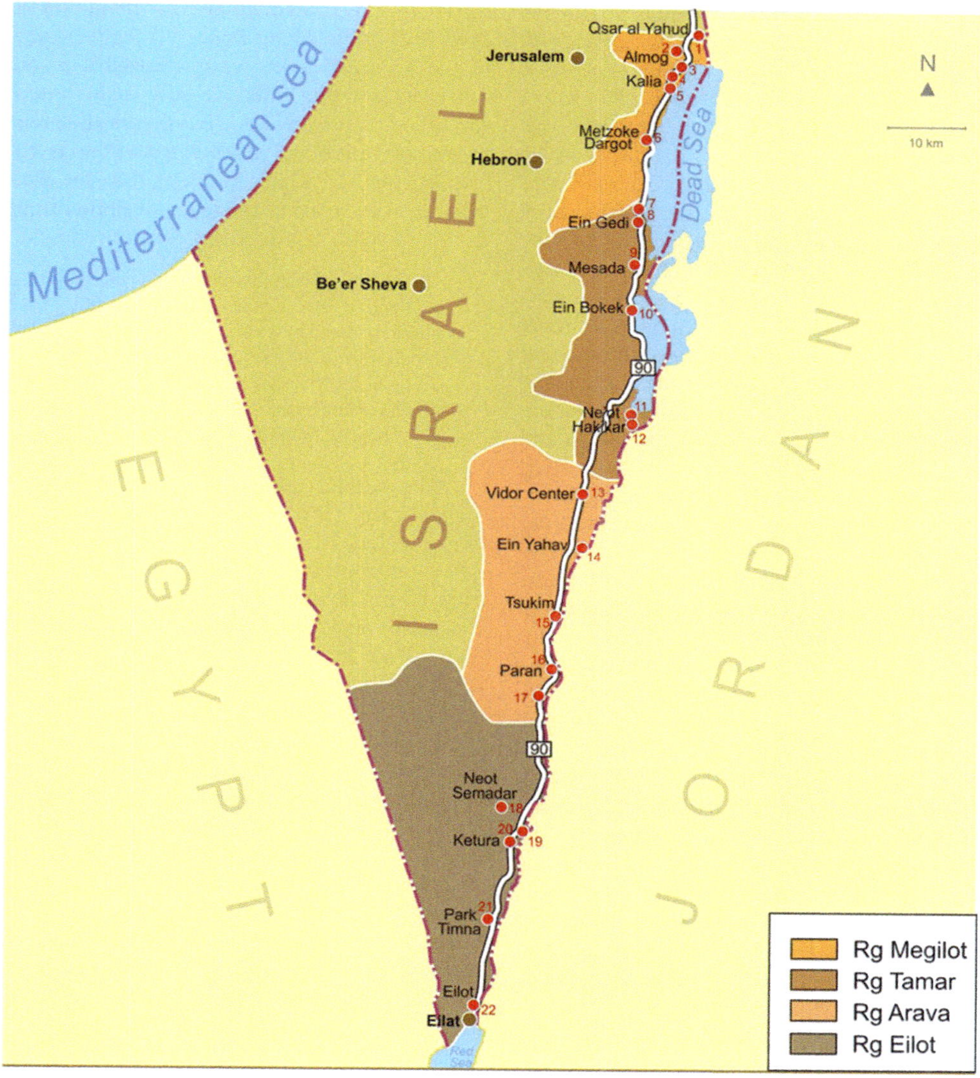

Numbers on the map indicate:

1. Qasr al-Yahud-National Park
2. Almog Vacation Village
3. Kalia Vacation Village
4. Qumran National Park
5. Einot Tsukim (Ein-Feshkha) Nature Reserve
6. Metzoke Dragot Vacation Village
7. Ein-Gedi Nature Reserve
8. Ein-Gedi Hotel
9. Masada National Park
10. Ein Bokek Hotels
11. Sodom Valley B&Bs
12. Sodom Valley B&Bs
13. Vidor Center – A Window to ARAVA Agriculture
14. Ein-Yahav B&Bs
15. Tzukim B&Bs
16. Paran B&Bs
17. The 101 km Rode Inn
18. Lotan Vacation Village
19. Ketura Vacation Village
20. Neot Semadar Vacation Village
21. Timna Park
22. Eilot Vacation Village

Fig. 2.1. Route 90 South, Israel.

groups of interviewees (government, regional councils and NGOs), it exposed a broad range of perspectives on how these groups might perceive a potential tourism route as a concept for inter-regional cooperation.

The author conducted all interviews in the interviewees' first language, Hebrew. The author, a native Hebrew speaker, translated all of the interviews from Hebrew to English and back into Hebrew to ensure validity of the verbiage. The interviews were analysed using Atlas software (Version 7), which enables users to insert large amounts of raw data while analysing and utilizing these data to enable understanding of hidden contexts, patterns and shared meaning among different interviewees (Mehmetoglu and Dann, 2003). The raw data analysis involved qualitative data coding to process each interview according to repetitive patterns in order to identify shared themes. Relevant quotes were extracted from these themes. Coding text is a method of reducing data while extracting prominent themes (Pan *et al.*, 2008). In this research, this analytical approach enabled the processing of large amounts of data across different interviewees, exploring and finding the challenges in tourism route establishment. It is important to mention that data are updated up to October 2021, using secondary and primary resources – regional council websites and conversations with tourism directors from the regional councils, respectively. This data update enabled the researcher to see that the situation witnessed during the 2016 data collection, that is to say no shared tourism route, is ongoing, making the 2016 data still very much relevant.

2.5 Findings

Throughout the interviews, the author discussed two main initiatives with each of the interviewees: 'Arts and Crafts Along Route 90 South' and 'Arava Way'. The first initiative took place in 2010 with the goal of facilitating collaboration between the four regional councils along Route 90 South. The name of the initiative suggests it is a tourism route, yet as all of the interviewees clearly explained, it amounted to a mere brochure (online and printed) and little, if any, awareness from the tourism businesses along the route. When asked to discuss the failure of 'Arts and Crafts Along Route 90 South', each of the 15 interviewees characterized it as an unclear marketing initiative that lacked interest and thus cooperation. It appears that this premature initiative suffered from low levels of participation, which indicates that the idea originated with one of the regional councils. As the Eilot Regional Council representative stated:

> Yes, I think I remember 'Arts and Crafts Along Route 90 South', it was Tamar's [regional council] idea. They wanted to produce a brochure with all the artists and craftsmen along Route 90 so we gave them the contact details of our businesses but it ended with only some printed materials, not much more.

The second initiative, 'Arava Way', only included the two regional councils from the Arava region. This route presents similar landscapes and agricultural villages throughout, yet they did not achieve cooperation and the route was finally deployed as a marketing platform for only one of the regional councils (Eilot – South Arava). As opposed to creating a tourism route based on geographic boundaries or demand, the initiators chose only to work within the pre-established administrative boundaries. It became clear from the interviews that each regional council works independently as a general rule with no shared tourism route. Although attempted or planned, cooperation between regional councils along the route is all but absent, whether for the Dead Sea and Arava regions together, or just for the Arava region.

Discussion around the 'Arava Way' allowed for a broader understanding of the explanation. First of all, the primary rationalization for separating the Arava and Dead Sea regions in terms of a tourism route amounted to significant differences in their identities, from geography to branding and economic scale of activity. It appears that the reluctance to establish one brand for the Dead Sea and Arava regions originated with the Arava region leaders, who decided they first needed to establish their own brand. As the Central Arava Regional Council representative explained:

> The Dead Sea brand is a very strong and familiar brand, it even won awards. The Dead Sea brand is associated with the hotel area in Ein-Bokek. We [the Arava region] are trying to establish

> different values for our brand and associate it with desert/rural tourism.

In a more detailed conversation about the 'Arava Way' project and why it lacks cooperation, both within the Arava region itself and along with the Dead Sea region, it became clear that it is due to disparate scales of economic activity. While the southern part is close to the city of Eilat (a major resort destination in Israel) and has various vacation villages functioning as one large unit, the central and northern parts work mainly with B&Bs and independent, small-scale attraction-providers far from any major city. Vis-à-vis the Dead Sea region's Ein-Bokek hotel area, the rift is even wider. Similar circumstances created a gap within the Dead Sea region itself between the southern part of Ein-Bokek and the northern part of Kalia where smaller-scale tourism takes place. On the whole, the differences in scales of economic activity are a significant concern. As the majority of NGOs stated and as the ICA Fund and Merage Foundation representatives described:

> Some regional councils are working on a larger scale of activity than others. It happens in the Dead Sea region as well as the Arava region. At the Dead Sea it is Tamar Regional Council who has many hotels and in the Arava region it is Eilot [Regional Council] who has large vacation villages. You see how hard it is to create cooperation between them when one always feels much less dominant than the other. (ICA Fund)
>
> I think the main obstacle is the Ein-Bokek hotel area [Dead Sea region]. It is much larger than other areas on the route, you know, in terms of economic activity. It makes it harder to cooperate when large hotel chains need to work with small rural businesses. (Merage Foundation)

It is important to mention that the hotel area of Ein-Bokek, located in the southern part of the Dead Sea region, also creates a kind of divide within the region itself between the Tamar and Megilot regional councils. The latter is responsible for the northern end of the Dead Sea where smaller-scale economic activity takes place, with no hotel chains and mainly vacation villages and B&Bs. Megilot Regional Council's leader exemplified, using the case of marketing campaigns, why regional councils operating at significantly different scales of economic activity – albeit sharing the same regional brand – find it difficult to cooperate:

> Every time there is a marketing campaign for the Dead Sea, all the attention goes to the hotels and we do not have hotels. But I know I can't complain because the Hotel Association and Tamar Regional Council are the ones who invest most of the money. Our scale of activity and budgets are much lower and so are our benefits.

Without exception, every interviewee mentioned the imbalance along the route. It appears that the 300 km route is marked by many different levels of development, with emphasis on the vibrant hotel area with 4000 rooms, restaurants, beaches, a new promenade and more (Ein-Bokek, Dead Sea). Moreover, the level of interest from the government, as reflected by the Ministry of Tourism and The Dead Sea Preservation Government Company, was very high when it came to the Ein-Bokek hotel area:

> We are focusing on preparing physical infrastructure in Ein-Bokek [Dead Sea hotel area] in order to double the amount of bedroom offerings, from 4300 to 9000 in 5 years' time. Our purpose is to make this area one of the most attractive hotel areas in Israel.

Furthermore, since bureaucratically it is easier to work with the Dead Sea hotel area in terms of certifications and building schemes, the government considers a shared tourism route with the rural areas a challenging task. As opposed to the hotels, many rural tourism businesses have only recently (in the last decade) moved from agriculture to tourism, with many of them adding accommodation to the premises of their farms. As reflected by the Ministry of Tourism representative: 'The most significant disadvantage for developing this route is that rural areas, which comprise most of the route, have no option to grow in terms of hospitality services due to lack of proper schemes [building authorization, standardization].'

The NGOs that work with the different regional councils provided another angle on collaborating with inexperienced business owners in small rural areas as opposed to hotel chains. As the Jewish National Foundation representative stated: 'Small businesses are busy with survival and it is hard to set standards. They mostly seek benefits from locality and cooperation with nearby businesses.'

In summary, it appears that the reasons for creating a separation between the regional councils along Route 90 South are: (i) branding; (ii) different scales of economic activity; and (iii) standardization. The following section discusses these factors in greater depth.

2.6 Discussion and Analysis

This section discusses how the senior stakeholders of Route 90 South, Israel (at the regional and national levels as well as those working in NGOs), perceive the challenges to establishing a new tourism route. Their responses are analysed with reference to the literature in order to gain new insights.

2.6.1 Branding

The Arava and Dead Sea are at different stages of brand awareness. While the Dead Sea is a well-established brand, known for inbound tourism, the Arava is an emerging brand mainly known for domestic tourism. Although the literature clearly shows that successful tourism routes may have various brands at different levels of awareness (Meyer, 2004; McCarron *et al.*, 2015), the majority of senior stakeholders in this study believe it is a significant obstacle. Moreover, the majority of senior stakeholders perceive a tourism route as an unsuitable product for the Dead Sea and Arava regions together. They point to a discrepancy in their geographical characteristics (different topographies and natural resources), which in turn results in different branding.

The literature shows that different branding, landscape and topography may prove advantageous for a tourism route project (Meyer, 2004; McCarron *et al.*, 2015). Going further, the literature clearly demonstrates success in diversified tourism routes, which link major brands with less distinctive brands (Hannam *et al.*, 2014; Rogerson, 2015). The newly formed tourism route brand may comprise several brands that stand alone yet exist in cooperation with the tourism route umbrella brand, adding more demand without sacrificing the existing brands. However, in this research the senior stakeholders focused on familiar branding (Dead Sea) in relation to unfamiliar branding (Arava) with an emphasis on the differences in brand values. While they consider the Dead Sea to have core brand values unique to resorts/hotels and various related services (spas, shopping malls, etc.), the Arava aims for the rural values of agricultural villages and desert activities (e.g. hiking).

Although the literature offers examples of successful tourism routes that span very different regions and services (e.g. Wild Atlantic Way; McCarron *et al.*, 2015), there is a lacuna in the literature when it comes to in-process branding. That is to say, what happens if one brand is in the process of formalization (Arava) while others are already well formalized (Dead Sea)? From this research it is clear that when in the process of creating a brand for one region alone, joining forces with another region for the creation of a shared tourism route brand becomes extremely challenging, perhaps impossible. This finding is new to the literature and sheds important light on the challenges that may arise before a tourism route is established. Interested parties must therefore attend to the challenge of brand formation by considering both the branding of each part of the route and how such challenges allow for or disallow a shared tourism route brand.

2.6.2 Different scales of economic activity

Along Route 90 South in the Dead Sea and Arava regions, there are several principal areas of activity. The largest, Ein-Bokek (14 hotels), is considered a threat to the others as it attracts most of the tourists and is naturally distinct from the rural tourism along the route. This perception was common among the majority of stakeholders and proved to be a significant obstacle to cooperation in a tourism route initiative. In the literature, however, successful tourism routes are usually those that connect varied tourism areas (Telfer, 2001). The main challenges discussed in the literature regarding disparate levels of development are a lack of road infrastructure and a lack of access to certain parts of the road, most common to developing countries (Hemme, 2005; Ahebwa *et al.*, 2016). In this research, which deals with a developed country, the road is equally accessible but provides access to locales with unequal scales of economic activity, hindering cooperation. Such a circumstance sets this research apart from any other studies of its kind to date. Moreover, in the literature, various case studies offer up examples of tourism routes with highly diversified economic activity. For example, 'The Wild Atlantic Way' in Ireland covers some 2500 km with cities, hotel areas, small rural villages and more, which showcases that different scales of economic activity may be a desirable attribute for a successful tourism route and not the opposite (McCarron *et al.*, 2015; Dunne *et al.*, 2021).

This raises a question regarding the stakeholders of Route 90 South and why they perceive different scales of activity as a threat rather than an opportunity. Where does this concern emanate from, and why is the Ein-Bokek hotel area such a substantial obstacle to cooperation? In search of an answer, this research considers a significant factor – the national focus on Ein-Bokek and plans to double the number of rooms within a couple of years. It appears that heavy investment in the infrastructure and marketing of the Ein-Bokek hotel area creates an extremely uneven scale of economic activity to the detriment of all other tourism offerings along the route. In the literature, while tourism routes are proven to encompass different scales of economic activity, no case study has dealt with a situation in which one area along the route is the focus of such substantial governmental investment in comparison to the others. In this case study, the spotlight is on an area that for all intents and purposes represents the Dead Sea region, presenting a great challenge to forming a shared tourism route with the Arava region.

2.6.3 Standardization

As mentioned previously, different levels of development exist among different areas along Route 90 South. This further results in some areas that are more standardized than others, with particular emphasis on the Ein-Bokek hotel area. The need to bridge between different levels of standards is therefore evident and was mentioned by all stakeholders. Since the literature offers no documentation of that which occurs before a route's establishment, there is no evidence of how different standards might affect willingness to cooperate or how they might be bridged. The literature does show, however, that standardization schemes (e.g. training, certification) exist and assist in the development of a tourism route initiative (Iles and Prideaux, 2011).

In this case study it is clear that senior stakeholders place great importance on the way these discrepant standards affect their ability to form a tourism route. The Arava region is considered to be less developed than the Dead Sea region, and within each region the southern portion is considered to be more standardized than the northern portion. Although the literature confirms that a tourism route project may help set standards or even raise them through certification schemes, guidance, support and synergy between a wide range of stakeholders, in this research not one of the governmental offices or NGOs stated they would be willing to lead a tourism route project. Each regional council tends to present its own vision so that its political tenure might become associated with change; this is also defined as 'political point scoring' (Sivijs, 2003, p. 344), which manifests in administrative borders. The bottom line of this situation is a 300 km linear route with four regional councils working separately, with each regional council working according to its own vision and proceeding according to the nature of their own hospitality services, standards, scale of activity and branding.

2.7 Conclusions

This research presents a new angle on tourism routes as a shared initiative for various stakeholders from the government, NGOs and local authorities in the pre-establishment phase of a route. Throughout the literature review and analysis, different tourism route initiatives are taken into consideration. In this research, similar to other case studies from the developed and developing world, it emerged that cooperation between stakeholders is by far the most challenging element (Hardy, 2003; Lourens, 2007; Rogerson, 2007;

Iles and Prideaux, 2011; Myles, 2014; Bregoli *et al.*, 2016). The various reasons cited in the literature mainly revolve around inappropriate physical infrastructure and lack of funding. The few that did investigate the more subtle elements report that a lack of trust may exist alongside concerns for an imbalance of power (Briedenhann and Wickens, 2004; Smith, 2011; Myles, 2014). This research goes into further depth to reveal the reasons for these concerns.

The findings indicate that a tourism route as a product, while designed to rise above administrative borders, is not always perceived as appropriate by the stakeholders responsible for its initiation. The reasons have largely to do with identity in the form of branding, scale of activity and standards. The more different they are, the harder it is to achieve cooperation. That being said, the literature shows how this very difference makes for a diversified and attractive tourism route (McCarron *et al.*, 2015). In this research, the competition between different portions of the route and an added lack of trust in the route as a whole result in an attempt by each local authority to brand itself, increase its activity and standardize it.

These results shed important light on the views of stakeholders, an easily overlooked factor in approaching a new tourism route initiative. The recommendations are therefore to analyse each part of the route as an individual unit before trying to connect them all as a whole, attending to these three aspects: brand status, scale of activity and local standards. It is very important to consider the variance that exists along the route, as this research clearly demonstrates that a high degree of variance may result in greater challenges to cooperation. In this research, the fact that one region is in the process of brand creation (Arava) and the other is a well-known brand (Dead Sea) has a tremendous impact on the ability to form a new tourism route brand for Route 90 South. As for the scale of activity, one area on the route (Ein-Bokek) is the locus of heavy government investment and boasts a much larger scale of activity when compared to any other part of the route. This fact created a situation where the connection with other areas on the route seemed too artificial and next to impossible (even within the Dead Sea region). This difference, although on a much smaller scale, is evident in the Arava region as well between the north-central part and the southern part; this hindered cooperation within the region itself.

It is therefore recommended to establish a clear understanding of how each part of the route not only competes with but also completes the others, even when working on different scales of activity. It is also recommended to clarify how different standards that result from different scales of activity might be overcome with training and certification schemes. This effective method, according to the literature (Iles and Prideaux, 2011), may work to bridge significant gaps in standards and moreover create synergy and empower different service providers. It is important to note that this research holds limitations as a case study of two distinct regions. There are likely to be different scenarios in which stakeholders will see the three aforementioned factors (branding, scale of activity, standards) not as limiting, but empowering. However, cultivating an awareness of these factors and broadening the perspective on how they may complete each other rather than compete could be of great value to the odds of a tourism route initiative's success. Further research should focus on private-sector stakeholders and challenges that arise from their point of view. Moreover, additional efforts to compare these findings with those of other tourism routes will strengthen the quality of future research.

References

Ahebwa, W.M., Aporu, J.P. and Nyakaana, J.B. (2016) Bridging community livelihoods and cultural conservation through tourism: case study of Kabaka heritage trail in Uganda. *Tourism and Hospitality Research* 16(2), 103–115. DOI: 10.1177/1467358415589659.

Beritelli, P. (2011) Cooperation among prominent actors in a tourist destination. *Annals of Tourism Research* 38(2), 607–629. DOI: 10.1016/j.annals.2010.11.015.

Berti, E. (2015) How to create a cultural route: project phases and criteria. In: Council of Europe. (ed.) *Cultural Routes Management: From Theory to Practice*. Council of Europe Publishing, Strasbourg, France, pp. 35–41.

Božić, S. and Tomić, N. (2016) Developing the cultural route evaluation mode (CREM) and its application on the trail of Roman Emperors, Serbia. *Tourism Management Perspectives* 17, 26–35. DOI: 10.1016/j.tmp.2015.11.002.

Bregoli, I., Hingley, M., Del Chiappa, G. and Sodano, V. (2016) Challenges in Italian wine routes: managing stakeholder networks. *Qualitative Market Research* 19(2), 204–224. DOI: 10.1108/QMR-02-2016-0008.

Briedenhann, J. and Wickens, E. (2004) Tourism routes as a tool for the economic development of rural areas–vibrant hope or impossible dream? *Tourism Management* 25(1), 71–79. DOI: 10.1016/S0261-5177(03)00063-3.

Byrd, E.T. (2007) Stakeholders in sustainable tourism development and their roles: applying stakeholder theory to sustainable tourism development. *Tourism Review* 62(2), 6–13. DOI: 10.1108/16605370780000309.

Carden, A.R. (2006) Sign of the times: an examination of the self-concept and emotional motivation of travelers on route 66 (1926–present). *Journal of Travel & Tourism Marketing* 20(2) 51–61.

Carson, D. and Cartan, G. (2011) Touring routes – types, successes and failures. In: Prideaux, B. and Carson, D. (eds) *Drive Tourism: Trends and Emerging Markets*. Routledge, Abingdon, UK, pp. 246–259.

Čeginskas, V.L.A., Kaasik-Krogerus, S., Lähdesmäki, T. and Mäkinen, K. (2021) Constructing social Europe through European cultural heritage. *European Societies* 23(4), 487–512. DOI: 10.1080/14616696.2021.1913508.

Damayanti, M., Scott, N. and Ruhanen, L. (2017) Coopetitive behaviours in an informal tourism economy. *Annals of Tourism Research* 65, 25–35. DOI: 10.1016/j.annals.2017.04.007.

Darabseh, F.M.Q. (2010) A strategy for the development of a tourist trail of the decapo is sites in northern Jordan. Doctoral dissertation, University of Birmingham, Birmingham, UK.

de Araujo, L.M. and Bramwell, B. (2002) Partnership and regional tourism in Brazil. *Annals of Tourism Research* 29(4), 1138–1164. DOI: 10.1016/S0160-7383(02)00033-6.

Dunne, G., Sprince, E. and Griffin, K. (2021) The wild Atlantic way: a tourism journey. *Irish Geography* 53(2), 127–144.

Freeman, R.E. (2010) *Strategic Management: a Stakeholder Approach*. Cambridge University Press, New York. DOI: 10.1017/CBO9781139192675.

Garfunkel, Z., Zak, I. and Freund, R. (1981) Active faulting in the dead sea rift. *Tectonophysics* 80(1–4), 1–26. DOI: 10.1016/0040-1951(81)90139-6.

Goussous, J. and Haddad, L.G. (2014) Innovation of new tourism trails and its effect on the ajloun touristic process. *Journal of American Science* 10(4s), 50–60.

Graf, M. and Popesku, J. (2016) Cultural routes as innovative tourism products and possibilities of their development. *International Journal of Cultural and Digital Tourism* 3(1), 24–44.

Hannam, K., Butler, G. and Paris, C.M. (2014) Developments and key issues in tourism mobilities. *Annals of Tourism Research* 44, 171–185. DOI: 10.1016/j.annals.2013.09.010.

Hardy, A. (2003) An investigation into the key factors necessary for the development of iconic touring routes. *Journal of Vacation Marketing* 9(4), 314–330. DOI: 10.1177/135676670300900402.

Hayes, D. and Macleod, N. (2008) Putting down routes: an examination of local government cultural policy shaping the development of heritage trails. *Managing Leisure* 13(2), 57–73. DOI: 10.1080/13606710801933420.

Hemme, D. (2005) Landscape, fairies and identity: experience on the backstage of the fairy tale route. *Journal of Tourism and Cultural Change* 3(2), 71–87. DOI: 10.1080/09669580508668488.

Iles, A. and Prideaux, B. (2011) The Savannah way: developing a successful touring route. In: Prideaux, B. and Carson, D. (eds) *Drive Tourism: Trends and Emerging Markets*. Routledge, Abingdon, UK, pp. 311–322. DOI: 10.4324/9780203880395.

King, S.A. and Davis Gatchet, R. (2018) Marking the past: civil rights tourism and the Mississippi Freedom Trail. *Southern Communication Journal* 83(2), 103–118. DOI: 10.1080/1041794X.2017.1404124.

Krakover, S. (2014) From tourism clusters to a tourism route along route 90 South, Israel (translated from Hebrew). *Negev, Dead Sea and Arava Studies* 6(1), 1–7.

Le Gouais, A., Panter, J.R., Cope, A., Powell, J.E., Bird, E.L. *et al.* (2021) A natural experimental study of new walking and cycling infrastructure across the United Kingdom: the Connect2 programme. *Journal of Transport & Health* 20, 100968. DOI: 10.1016/j.jth.2020.100968.

Lois-González, R.C. and Santos, X.M. (2015) Tourists and pilgrims on their way to Santiago. Motives, caminos and final destinations. *Journal of Tourism and Cultural Change* 13(2), 149–164. DOI: 10.1080/14766825.2014.918985.

López, L. and Lois-González, R.C. (2020) New tourism dynamics along the way of St. James. From undertourism and overtourism to the post-COVID-19 era. In: Pons, G.X., Blanco-Romero, A., Navalón-García, R., Troitiño-Torralba, L. and Blázquez-Salom, M. (eds) *Sostenibilidad Turística: Overtourism vs Undertourism*. pp. 541–552.

Lourens, M. (2007) Route tourism: a roadmap for successful destinations and local economic development. *Development Southern Africa* 24(3), 475–490. DOI: 10.1080/03768350701445574.

MacLeod, N. (2016) Self-guided trails – a route to more responsible tourism? *Tourism Recreation Research* 41(2), 134–144. DOI: 10.1080/02508281.2016.1147222.

Marschall, S. (2012) Sustainable heritage tourism: the Inanda heritage route and the 2010 FIFA World Cup. *Journal of Sustainable Tourism* 20(5), 721–736. DOI: 10.1080/09669582.2011.638385.

McCarron, C., McCusker, P. and Ó hAnnracháin, C. (2015) The Wild Atlantic Way as a catalyst for tourism and sustainable employment development in the North West of Ireland. Paper delivered at Tourism and Hospitality Research in Ireland Conference, Letterkenny Institute of Technology, 11–12 June 2015.

Mehmetoglu, M. and Dann, G. (2003) Atlas/ti and content/semiotic analysis in tourism research. *Tourism Analysis* 8(1), 1–13. DOI: 10.3727/108354203108750120.

Meyer, D. (2004) *Tourism Routes and Gateways: Key Issues for the Development of Tourism Routes and Gateways and Their Potential for Pro-Poor-Tourism*. Overseas Development Institute, London.

Moulin, C. and Boniface, P. (2001) Routeing heritage for tourism: making heritage and cultural tourism networks for socio-economic development. *International Journal of Heritage Studies* 7(3), 237–248. DOI: 10.1080/13527250120079411.

Murray, M. and Graham, B. (1997) Exploring the dialectics of route-based tourism: the Camino de Santiago. *Tourism Management* 18(8), 513–524. DOI: 10.1016/S0261-5177(97)00075-7.

Myles, P.B. (2014) Coastal route tourism: a vehicle for collaborative economic development in the eastern Cape, South Africa. *Tourism in Marine Environments* 9(3), 169–179. DOI: 10.3727/154427313X138 18453739512.

Olsen, M. (2003) Tourism themed routes: a Queensland perspective. *Journal of Vacation Marketing* 9(4), 331–341. DOI: 10.1177/135676670300900403.

Pan, S., Chon, K. and Song, H. (2008) Visualizing tourism trends: a combination of ATLAS.ti and BiPlot. *Journal of Travel Research* 46(3), 339–348. DOI: 10.1177/0047287507308318.

Prideaux, B. and Coghlan, A. (2011) Driving the desert: profiling four-wheel drive visitors. In: Prideaux, B. and Carson, D. (eds) *Drive Tourism: Trends and Emerging Markets*. Routledge, Abingdon, UK, pp. 246–259.

Rogers, A.G. and Leung, Y.F. (2020) Smarter long-distance hike: how smartphones shape information use and spatial decisions on the Appalachian trail. *International Journal of Wilderness* 26(2), 88–103.

Rogerson, C.M. (2007) Tourism routes as vehicles for local economic development in South Africa: the example of the Magaliesberg Meander. *Urban Forum* 18(2), 49–68. DOI: 10.1007/s12132-007-9006-5.

Rogerson, C.M. (2015) Unpacking business tourism mobilities in sub-Saharan Africa. *Current Issues in Tourism* 18(1), 44–56. DOI: 10.1080/13683500.2014.898619.

Shunnaq, M., Schwab, W.A. and Reid, M.F. (2008) Community development using a sustainable tourism strategy: a case study of the Jordan River Valley tourist way. *International Journal of Tourism Research* 10(1), 1–14. DOI: 10.1002/jtr.620.

Sivijs, A. (2003) The drive tourism program 2000-2002: project partners in the fast lane. *Journal of Vacation Marketing* 9(4), 342–353. DOI: 10.1177/135676670300900404.

Smith, N. (2011) Barriers to and benefits of clustering in drive tourism markets. In: Prideaux, B. and Carson, D. (eds) *Drive Tourism: Trends and Emerging Markets*. Routledge, Abingdon, UK, pp. 323–338.

Stoffelen, A. and Vanneste, D. (2016) Institutional (dis)integration and regional development implications of whisky tourism in Speyside, Scotland. *Scandinavian Journal of Hospitality and Tourism* 16(1), 42–60. DOI: 10.1080/15022250.2015.1062416.

Telfer, D.J. (2001) From a wine tourism village to a regional wine route: an investigation of the competitive advantage of embedded clusters in Niagara, Canada. *Tourism Recreation Research* 26(2), 23–33. DOI: 10.1080/02508281.2001.11081340.

Timothy, D.J. and Boyd, S.W. (2015) *Tourism and Trails: Cultural, Ecological and Management Issues*. Channel View Publications, Bristol, UK. DOI: 10.21832/9781845414795.

Viljoen, J., Viljoen, F. and Struwig, J. (2010) Pro-poor tourism routes: the open Africa experience. *Acta Academica* 42(4), 65–90.

Wang, Y. and Krakover, S. (2008) Destination marketing: competition, cooperation or coopetition? *International Journal of Contemporary Hospitality Management* 20(2) 126–141. DOI: 10.1108/09596110810852122.

Werner, C. (2003) The new silk road: mediators and tourism development in Central Asia. *Ethnology* 42(2), 141. DOI: 10.2307/3773779.

Zoomers, A. (2008) Global travelling along the Inca route: is international tourism beneficial for local development? *European Planning Studies* 16(7), 971–983. DOI: 10.1080/09654310802163769.

3 Women's Empowerment through Community-based Tourism in Bethlehem Governorate, Palestine: An Exploratory Study

Rami K. Isaac[1]* and Wisam Abbasi[2]
[1]*Research Group Tourism Impacts on Society, Breda University of Applied Sciences, Breda, The Netherlands;* [2]*Japan International Cooperation Agency, Tourism Development Advisor, Palestine*

3.1 Introduction

Numerous scholars stated that gender and tourism research have gained momentum because of the need to better understand the potential, opportunities and pitfalls confronting tourism stakeholders regarding poverty reduction (Tucker and Boonabaana, 2012; Alarcón and Cole, 2019); policy making and projects (Ferguson, 2010a, b; Ferguson and Alarcón, 2015), and decent working conditions (Huete *et al.*, 2016).

The tourism sector plays a crucial role with the purpose of decreasing gender inequalities by allowing women to participate in economic activities, and gain access to resources and opportunities for jobs in this sector (UNWTO, 2011). However, it is widely acknowledged that empirical evidence linking women's participation in tourism to female empowerment is scarce (Movono and Dahles, 2017). As Perkins and Zimmerman (1995, p. 571) state, empowerment has 'become a vital construct for understanding the development of individuals, organizations and communities'. Its roots can be traced to the initial work of Boserup (1970). The key study of Boserup (1970) concluded (at that time) that women should be included in the developmental agenda as women are marginalized in every way of life due to male-biased society, little freedom of voice and low salary.

Klasen (2002) further highlighted that gender inequality largely harms the country's economic development and leads to negative impacts in the long term (Klasen and Lamanna, 2009). Empowerment refers to the development of self-determination of choice and action. This applies to women as well as to other disadvantaged or socially excluded groups, but women's empowerment includes some unique additional elements as argued by Malhotra *et al.* (2002). Women's empowerment requires systemic transformation in not just institutions, but fundamentally in societies supporting patriarchal structures.

Community-based tourism (CBT) has become a commonly used approach in marginalized communities, mainly because of its potential for bottom-up development and control over

*Corresponding author: Isaac.r@buas.nl

 Tourism Planning and Development in the Middle East (eds S. Kladou *et al.*)
DOI: 10.1079/9781789249149.0003

tourism planning and management (Bello *et al.*, 2017). CBT strengthens social and economic capacities. Tourism can affect community development: it can encourage communities to interact in development interventions by empowering them (Sofield, 2003).

CBT is thought to contribute to poverty reduction, because it has been demonstrated to support job generation (López-Guzmán *et al.*, 2011) as well as improve localized economic connections resulting in reduced economic leakages (Lapeyre, 2010). CBT is also viewed as a positive driver as it encourages local participation, empowerment and decision making (Duffy *et al.*, 2015). The participation of women, minorities and native people in CBT initiatives is vital for achieving economic, environmental and social development (Harris *et al.*, 2001).

Al-Oun and Al-Homoud (2008, p. 38; see also Daher, 2000) in their study in Jordan state that:

> new approaches to defy traditions of top-down planning follow extensive debates in recent years on the need for more sustainable forms of tourism development, which recognise the need for policies designed to use less sophisticated infrastructures and support local community involvement and empowerment in tourism with its social, cultural and financial rewards.

These issues are particularly relevant in developing countries among populations that traditionally have been marginalized in society, such as ethnic minorities, indigenous people and women (Johnston, 2003).

While tourism is one among the foremost important industries in Palestine, it is underrepresented within the academic literature (Isaac *et al.*, 2016). Previous studies have analysed the trends, profiles and motivations of mass tourists (pilgrimage) and image formation in Palestine (Isaac, 2009, 2010a, b; Çakmak and Isaac, 2012; Isaac and Ashworth, 2012; Isaac and Eid, 2019). However, studies on gender and women empowerment have been largely ignored. This chapter presents a case study of CBT projects in Bethlehem, as this form of tourism is gaining ground in Palestine in general and Bethlehem in particular (Isaac and Eid, 2019) as an 'add on' to the dominant pilgrimage form of tourism. Therefore, this study aims to examine the factors responsible for women's low participation in CBT in Bethlehem, identify the challenges for women's participation and highlight some of the initiatives that could be taken to enhance women's participation and lead to their empowerment through tourism. To the best of the author's knowledge, no study has so far been published with a focus on women's empowerment in Palestine. This chapter therefore closes the existing gap in the literature and provides valuable knowledge on women's empowerment through CBT in Bethlehem.

3.2 Tourism Planning and Development and the Role of Women

As Movono and Dahles (2017, p. 683) state, 'despite the overwhelming evidence that women are important producers of goods and services in the tourism industry, female involvement in tourism businesses has always been an understudied area'. It is agreed that tourism development enhances income opportunities for women (Sinclair, 1997; Boonabaana, 2014). It empowers women to gain increased income and, in some cases, adjusts dominant gender norms and traditional power relations (Castelberg-Koulma, 1991). Employment opportunities for women outside of their homes and traditional roles have generated supplementary income for women, improving their living standards and reducing their economic dependence on men (Cave and Kilic, 2010; Cherayi and Jose, 2016). In addition to enhanced individual and household economic status, participation in tourism also can improve women's welfare and academic achievement (Knight and Cottrell, 2016; Arroyo *et al.*, 2019). However, female employment in tourism, as in other sectors of the economy, is organized along gender lines and follows dominant gender norms (Garcia-Ramon *et al.*, 1995; Cukier *et al.*, 1996). It usually happens in spaces defined as 'women's work' and mirrors women's domestic caring and mothering roles, such as cooking, cleaning, washing clothes, and providing services such as hairstyling and manicures (Gentry, 2007).

The subsequent section will introduce women's empowerment and discuss different models and frame works that have been used in the literature according to which empowerment in tourism can be categorized.

3.3 The Empowerment of Women

Gender, equality and empowerment of women are among the United Nations' Millennium Development Goals (MDGs) and Sustainable Development Goals (SDGs) (Malema and Naidoo, 2017; O'Hara and Clement, 2018). Though much advancement has been made, difficulties such as gendered violence (Malema and Naidoo, 2017), the feminization of poverty (Bárcena-Martín and Moro-Egido, 2013), biased depiction of women (Maruyama and Woosnam, 2020) and little access to education (Cherayi and Jose, 2016) persist to encumber advancement in the empowerment of women (Bárcena-Martín and Moro-Egido, 2013; Malema and Naidoo, 2017).

This is predominantly evident in developing countries (Boserup, 1970; Laszlo *et al.*, 2020). Empowering women must be primarily focused on the household level since the household and interfamilial relations are a central venue of women's disempowerment. Duflo (2012) debated women's empowerment and economic development through the tourism sector and claimed that these two terms are closely related.

Empowerment is a multilayered and deeply contextual concept with economic, social, psychological, educational and political dimensions (Rowlands, 1995; Scheyvens, 1999; O'Hara and Clement, 2018). These five dimensions exhibit different representations at different scales from the self, household and community to the broader society (Scheyvens, 1999; Cherayi and Jose, 2016; O'Hara and Clement, 2018). Although there is no universal and agreed definition of empowerment, it is widely agreed that autonomy, independence, access to opportunities, prospects and capital, and freedom to make one's own decisions are essential to the concept of empowerment (Scheyvens, 1999, 2000; Bishop and Bowman, 2014; Moswete and Lacey, 2015). The current study draws from the relevant tourism-based literature to conceptualize and define the concept of empowerment in the context of women's participation in CBT in Bethlehem.

Several models have been suggested for understanding the multifaceted nature of women's empowerment. For example, Scheyvens (1999) provides a pivotal work and sets the frame works with which empowerment in tourism can be discovered and categorized in specific dimensions. Scheyvens (1999) emphasizes the importance of local groups having some control over and sharing the advantages of tourism development at the local level (Scheyvens, 1999, p. 249). She suggested that empowerment comprises four dimensions: economic, political, social and psychological.

Economic empowerment refers to the upgrading of women's access to and control over resources for production, such as land and capital, access to employment and business chances, and economic competence in and outside of the home (Scheyvens, 1999, 2000; Cherayi and Jose, 2016; Laszlo *et al.*, 2020). Political empowerment largely includes political representation, the inclusion of women in positions of power, and access to decision making for oneself, one's family, the community and wider society (Scheyvens, 1999; Cherayi and Jose, 2016). Social empowerment involves a complex range of aspects, including improved family and social status, social networks and relations, safety in family and society, equality, self-determination over marriage, social inclusion and enhancements in social institutions (Scheyvens, 1999, 2000; Cherayi and Jose, 2016; Joo *et al.*, 2020).

In addition to the economic, political and social dimensions mentioned above, empowerment 'must also include the processes that lead people to perceive themselves as able and entitled to occupy that decision-making space' (Rowlands, 1995, p. 87). Accordingly, psychological empowerment, highlighting the 'power within', includes self-confidence, critical awareness, respect of self and from others, fulfilment, leadership and creativity (Rowlands, 1995; Scheyvens, 1999; O'Hara and Clement, 2018; Joo *et al.*, 2020). In addition, Annes and Wright's (2015) study focused on empowerment as an exercise in the expansion of agency a *process of power to, power with and power within*. According to Su *et al.* (2020), an absence of psychological empowerment would hamper the realization of the other empowerment dimensions.

The next section will present how the CBT concept can be used as a tool for women's empowerment. CBT has for some time been argued as the only truly sustainable approach in developing an industry that relies on communities and their resources. It comprises community empowerment, ownership and management of the unique characteristics that make up a community's essence of place and function as a tourist attraction (Al-Oun and Al-Homoud, 2008).

3.4 CBT as a Route for Empowerment

CBT is repeatedly labelled as an alternative tourism approach designed to combat mass tourism in the developing world and in addition to benefit communities through grassroots development, local involvement, empowerment and capacity building (Dangi and Jamal, 2016).

There is no consensus on the definition in the academia about CBT, but the characteristics of CBT can be summarized as local ownership of projects, strong participation and involvement of local communities in all stages of project development, and cross-cultural interaction between host and guest (Goodwin and Santilli, 2009). The benefits of CBT to the communities include economic benefits, such as employment and eventually poverty reduction (Müller *et al.*, 2020) and empowerment of communities (Salazar, 2012).

Belsky's (1999, p. 653) research in south-eastern Belize verified that women who owned bed and breakfast establishments remained in control of the income they generated because operating such businesses was not viewed as 'work', while research by Dahles and Bras (1999, p. 35) displayed that female entrepreneurs' elevated income often leads to 'increases in social status, empowerment, and the creation of new occupational opportunities'. Additional benefits of tourism employment may include decreased birth rates, increased economic self-sufficiency (Kousis, 1989; Chant, 1992), increased status and decision-making power within the household and possibly in the public sphere (Sinclair, 1997, p. 232).

Nevertheless, Ferguson and Alarcón (2015) emphasize that there are certain aspects of gender inequality that are specific to CBT. The cause of such inequality is the difficulty to integrate gender analysis results into the sustainable tourism paradigm. Their study showed limitations to the gender integration approach in CBT. They discuss ways to operationalize gender integration in CBT. Their proposal takes into consideration three important categorized areas: (i) promotion of equal opportunities between women and men and women's active participation in decision-making processes; (ii) training and decent work that challenges gender roles and stereotypes; and (iii) greater equality between women and men within the workplace, household and community.

CBT as a route for empowerment means active participation, local ownership, control and power over tourism decision-making, resources and lives (Lapeyre, 2010). As Dolezal and Novelli (2020, p. 3) state, CBT determines the difference between tourism in the community and CBT and enables the community to perform as the 'subject' rather than the 'object' of CBT.

CBT and empowerment are principally understood and valued for their economic dimension (Connell and Rugendyke, 2008). Accordingly, many often ignore 'psychological empowerment' from feelings of self-esteem and pride, 'social empowerment' from community cohesion and sense of community, and 'political empowerment' from decision-making capabilities and opportunities (Scheyvens, 2002).

3.5 Context of the Study – Southern Governorate (Bethlehem)

The state of Palestine has a strategic location in the centre of the Middle East (see Fig. 3.1). In addition, it is one of the world's most important historical, religious and cultural sites, given its role in the three foremost monotheistic religions, Christianity, Islam and Judaism (Isaac *et al.*, 2016; Isaac, 2018). Pilgrimage is the backbone of the Palestinian economy.

Over the years, tourism has enhanced the Palestinian economy, especially in religious spots that depend on this sector, such as Bethlehem,

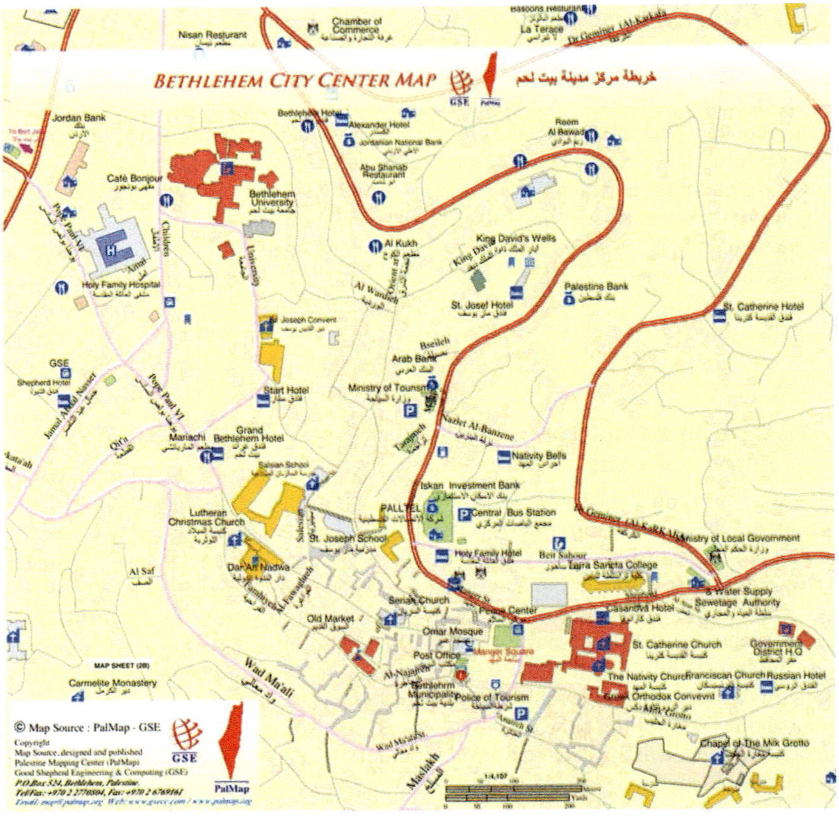

Fig. 3.1. Map of Bethlehem. Used with permission from copyright GSE Good Shepherd Engineering.

Jericho and East Jerusalem. Nonetheless, the political instability in the area has affected tourism negatively (Isaac, 2009; Isaac *et al.*, 2016). In 1948, Israel was established on almost 77% (Prior, 1997; Ibrahim, 2013; Isaac *et al.*, 2016) of the historical land of Palestine after a destructive war in the region.

In 1967, Israel occupied the remaining part of Palestine, which was divided into the West Bank (including East Jerusalem) and the Gaza Strip. Since 1967, the West Bank has been under military Israeli occupation. These political circumstances have affected all the industries in the area and especially tourism (Isaac *et al.*, 2016). Nevertheless, a new report, produced by the United Nations World Tourism Organization (UNWTO), confirmed that the world's fastest growing tourist destination for 2017 is Palestine (Nabbout, 2017). Extant studies suggest that the development of tourism is a cornerstone for the overall development of Palestine and its economy (e.g. Isaac, 2009; Isaac and Platenkamp, 2010; Isaac *et al.*, 2016; Isaac and Eid, 2019).

However, the economic disparity is linked to the Israeli's tourism monopoly and the struggle over tourism in Palestine (Berkowitz, 1997; Cohen-Hattab, 2004; Isaac, 2010a, b; Smith, 2010). Their works show how Zionists were seeking to establish control over the tourism sector in two ways: First, by creating a new form of Zionist tourism for western Jews. Second, by taking over the pre-existing and Palestinian controlled tourism religious and archaeological sites and marketing them as Israeli ones (Whitelam, 2013). These Israeli manoeuvres were used to manipulate the pilgrimage market and gain control of the tourism sector, allowing Palestinians only a small share of the pie (Isaac, 2010a, b).

As Cohen-Hattab and Shoval (2015, p. 65) state:

> the Zionist movement, defining tourism in political terms, consciously set out to exploit the tourist industry as part of its national political struggle. As such it launched a coordinated attack on the country's oral, visual, and written tourism media. Guidebooks, tourists' maps, advertisement, films and tour guides were all used to shape and present the desired Zionist take on Palestine.

Over the years, Israel simply prevented the Palestinians from benefiting from the tourism sector. Even after 1967, the tourism sector, especially the hotel industry, continued to suffer (Baumann, 2016), due to the Israeli financial constraints on this industry, which left it underdeveloped (Organisation of Islamic Cooperation, 2014; Arafeh, 2016) compared to the Israeli hotel industry.

The Palestinian society is known to be a patriarchal one. As stated by Haj-Yahia (2005), the patriarchal structure of the Palestinian society and gender inequality are both deeply rooted. In fact, the patriarchal society could be a major factor in the gender gap in Palestine. In relation to tourism, the gender gap is articulated in the statistics of the Palestinian Central Bureau for Statistics (PCBS) and shows that the number of employed individuals in the tourism sector in 2017 was 3100 females and 34,600 males (PCBS, 2017). The number of the working force in the tourism sector is thus 8% female and the ratio of women to men is 1 to 11.5 (PCBS, 2017).

The national export strategy of Palestine states that, in relation to the tourism export strategy 2014–2018, the tourism sector contribution to the national wealth has been increasing steadily since 2005, and is around 14% of gross domestic product (GDP). The number of Palestinians employed in this sector is significant. However, Palestinian women's contribution is less and mainly dominant in specific subsectors, such as the handcraft industry.

3.6 Women in Palestine

Ernudd (2007) argues that Palestinian women face two interconnected systems of oppression, namely occupation and patriarchy. This deterioration is harming the Palestinian women more than men, due to women's inferior status in the Palestinian society (Ernudd, 2007). The occupation (Isaac, 2013) combined with the patriarchal structures of the Palestinian Authority (PA) has served to further delimit the progress and development of Palestinian women.

Another important factor influencing Palestinian women's status is Islamic Fundamentalism in Palestinian society, which is a boundary to treaties and conventions on eliminating forms of discrimination and inequality towards women (Charlesworth and Chinkin, 2000).

In relation to Islamic Fundamentalism, Hamas and Islamic movements have dramatically impacted the civil society and women's position in Palestinian society (Abdo, 1999). To understand the status of Palestinian women, we need to underline two time periods related to Palestine: before and after the Oslo Peace process. The first one is the period before the Oslo Accords and the establishment of the PA in 1993 (Isaac, 2010a). The major change relates to the PA being recognized as the governing body, although the state remains economically, politically and geographically dependent on and occupied by Israel. As a result, there is no true sovereignty at this time (Abdo, 1999).

Another factor that contributes to the status of the Palestinian women is the lack of commitment from the PA to gender issues, because other major internal and external challenges have been prioritized. Consequently, gender advancement in Palestine is minimal due to the conservative society and the structure of the PA (Abdo, 1999). Accordingly, the relationship between men and women is not based on gender equality but is part of the dynamics surrounding the social relations, which is directly impacted by Israeli occupation policies (Haj, 1992; Isaac *et al.*, 2016).

There is a clear understanding of women's empowerment in Palestinian organizations because these organizations are exposed continuously to international discourse that focuses on gender balance (Kuttab, 2010).

Empowerment is incorporated into the policies related to grants and development projects funded by donor countries. The projects focus on three main types of empowerment:

political, economic and legal. Yet empowerment by international organizations implementing programmes in Palestine works on individual rather than collective empowerment. For example, Kuttab (2010) concludes that true empowerment will only be obtained if women have full rights as equal citizens. He adds that women should have decent and protected employment opportunities and not just access to work.

This kind of empowerment cannot happen under settler colonial occupation and patriarchal domination. Women's empowerment in Palestine is much more complicated and problematic since Palestinians are still living under occupation and this will hinder women's independence, equality and social justice (Kuttab, 2010). Kuttab's conclusions are the core of the process, especially when it comes to collective empowerment, the colonial occupation and its direct effect on true rather than superficial empowerment in Palestine.

3.7 Methodology

This qualitative research used purposeful sampling to include individuals who are knowledgeable on the research topic. The sample included individuals from NGOs, the private sector, the government and women from the community-based tourism sector in order to get various perspectives from these different stakeholders. In total, 30 semi-structured interviews were conducted (15 females and 15 males). The male interviewees' viewpoint was sought as they manage CBT projects. In addition to these interviews, the authors held interviews with five managers of current/ongoing CBT projects that focus on women's empowerment. Each interview lasted between 25 and 40 min.

The questions were divided into five sections. Section one, about empowerment, had four specific sections about economic, social, psychological and political empowerment of women (Scheyvens, 2002). The qualitative questions aimed to understand the perspective of different stakeholders in tourism from a gender perspective. All interviews were transcribed (see Table 3.1), and the data from the interviews were coded and analysed for explanatory themes, focusing on the proposed model by Scheyvens (1999), which has informed the way we have structured the findings below.

Table 3.1. List of interviewees, Bethlehem Governorate.

(MB) Male employee	Anonymous museum
(MU) male director	UNESCO site
CBT manager	Anonymous (CBT)
CBT manager	Anonymous (CBT)
CBT officer	Anonymous (CBT)
CBT manager	Anonymous Palestine Project (CBT)
(MOTA) 2 females working at MOTA	Palestine Ministry of Tourism and Antiquities
CBT manager & 5 females (FC) working in CBT	Battir village, CBT
8 females (FT) & 13 males (MT) working in the tourism sector	Anonymous

UNESCO, United Nations Educational, Scientific and Cultural Organization.

3.8 Findings

3.8.1 General understanding of women's empowerment

The research first allows interviewees to define, in their own words, women's empowerment. This step helps examine their own level of understanding. Results of the sample referred to the four dimensions related to empowerment (i.e. political, social, economic and psychological), and indicated that the women interviewed defined empowerment holistically. Most female respondents understood women's empowerment only in line with the economic dimension; a few of them understood it from a social dimension; and others understood it from the political dimension. As for their male counterparts, empowerment meant for the majority of them economic empowerment, others saw it as holistic empowerment and a few stated social and political aspects.

Almost half of the women interviewed saw themselves as holistically empowered whereas one-third of women identified as psychologically empowered; a few women perceived themselves as politically empowered; other women perceived themselves as economically empowered; and lastly, a few women perceived themselves as socially empowered. The same questions were asked of the male respondents and they saw women becoming empowered economically as a major dimension. More than half of the men perceived women to be empowered economically and three of them perceived women to be empowered holistically. A few of them thought that women are becoming empowered politically, psychologically and socially.

A question was presented to the interviewees regarding their perception of the effect of the unstable political environment on women's individual empowerment and the results were as follows: 15 women perceived that political instability has a negative impact on their empowerment as a whole, whereas a few of them believed that political instability affected the economic dimension of their empowerment. Other female respondents perceived that the political situation affected the social dimension of empowerment, and a few of them believed that political instability affected the political dimension. The majority of the male respondents perceived that political instability had a negative impact on women's empowerment; some of them believed that political instability affected the economic dimension of women's empowerment; others perceived that the political situation affected the social dimension; and a few of them believed that political instability affected the political dimension negatively. Finally, others believed that political instability had no impact on women's empowerment.

The interviewees were also asked how the low percentage of female participation in the tourism sector could be increased. Almost half of the female respondents believed that if women were holistically empowered in the four dimensions there would be a cause–effect relationship on the number of women working in the tourism sector. The ideas evolved around capacity building, entrepreneurship policies and projects tackling empowerment. On the contrary, the majority believed that if women had economic power this could result in more women's enterprises in the field of tourism and with that the number of women working in the field could increase. A few of them had no opinion or suggestions to answer this question. The male respondents thought that the economic dimension could have a better effect on increasing women's participation in tourism and others thought that holistic initiatives could increase women's participation in the tourism sector.

3.8.2 Economic dimension of women's empowerment

Women were asked to describe their economic status to understand their level of economic empowerment. The majority of women indicated that they earned a medium income, between 3200 and 6000 NIS (1 NIS is equal to 0.25 Euro cents) and a few indicated that they had a low income, less than 3200 NIS. The men were reluctant to describe their economic situation: the majority of them did not provide the information and a few described their economic situation as average.

One of the male respondents did not answer the question about 'property ownership' and the overwhelming majority showed that their property ownership is much better than that of their female counterparts; only a few men indicated that they had no property at all. Few women owned either a car or a house and most of the women did not have any property in their name.

All female respondents stated about inheritance and ownership that they didn't have any; their husbands owned the land. One of the women working in Battir village stated: 'Ownership is critical if we want women to have decision in using the land for tourism purposes[; currently] even if women are convinced to do projects in tourism, they need approval of their husbands.'

When women were asked to describe their personal economic journey, most of the women stated that they were employed and a few of them were participating in CBT projects seasonally. Those working seasonally or on an ad hoc basis do not have consistency or sustainability in their economic situation. Economic development is relatively low and most of the time women are employed and are not business owners. Economic development can increase women's profiles, but it is not enough to bring about complete equality between men and women. Most of the male respondents were employed in NGOs or in the public sector (MOTA). Some worked in CBT while others were entrepreneurs and had their own business or worked as freelancers in tour guiding and CBT.

One of the women in Battir village stated: 'We only need to increase our sales and work on marketing our products to be empowered economically, this is what we need, other empowerment dimensions are fine, we don't need anything else.'

Many women do not understand that economic empowerment will allow them to think beyond immediate daily survival in order to have greater control over both resources and life choices. They do not understand the real meaning of holistic empowerment. Thus, again the finding showed that women have a superficial empowerment built on male dominated decisions.

3.8.3 Social dimension of women's empowerment

When women were asked about their participation in local institutions such as women's associations or social institutions, almost half of them said that they participated in such associations. Those who did not participate said there was no time or that women did not have the opportunity to be part of these institutions. Men showed that they participated in local institutions at the same rate as women but a few of them did not answer and others showed that they participated in local institutions and supported women's participation.

As for the role of family support in women's participation in the tourism sector, all the interviewed women showed that their husbands and direct family provided full support and acceptance for women's participation in tourism activities. Nevertheless, they indicated that society did not provide the same support or acceptance. Additionally, when asked about other women's participation and social status, women seemed to be excluded from the social decision-making process within the local communities, which is an indicator of weak social empowerment.

3.8.4 Psychological dimension of women's empowerment

When women were asked whether they receive external psychological support, only a few of them said that they had such support, with the majority indicating that they lacked psychological support. A few didn't understand the term, while some didn't answer at all. Based on the analysis for those females employed in CBT, it can be concluded that tourism affects their self-esteem and confidence, which is a positive indicator that working in tourism empowers women psychologically regardless of the external support. Almost half of the men thought that women lacked psychological support; a few of them thought that women had such support; and some didn't understand the term.

One woman stated; 'we participate in all activities and training sessions provided by NGOs and funded by donors, but no project really tackles our needs and problems or sees things from our point of view. They come and do training for a week and then we see nothing - no sustainability or continuity.'

Whereas another woman stated: 'Also, organizations will pick certain women or women['s] associations known to them and ignore us and ignore other marginalized women and eventually ignore our needs, so equal opportunities is an important issue.'

Other women talked about the 'insecurity [of] home-based jobs in the tourism field. They are not registered and [there is] no rule to protect their rights or maintain economic security for them.'

This is an indicator that women do not feel psychologically empowered to start initiatives concerned with women's struggles to meet their needs for equality.

3.8.5 Political dimension of women's empowerment

Next, we asked women if they participated in positions that enabled them to be decision makers. Answers revealed that only one-third had political influence, in terms of the inclusion of women in positions of power and in the decision-making process. Almost half had none, while a few of them did not answer at all. Male respondents showed that one-third of men did participate, but did not have the power in decision making and a few didn't participate, while others did not answer at all.

One of the women stated, 'We have political empowerment based on the fact that we have women in the municipality working as employees. Unfortunately, as you already know, women are only fronts and not considered [as] decision makers even if they are employees.'

A question was presented to the interviewees regarding their perception of the effect on the unstable political environment on women's individual empowerment and the results were as follows. All female respondents perceived that the political instability had a negative impact on their empowerment as a whole, whereas some of them believed that the political instability particularly affected the economic dimension of their empowerment. Three female respondents perceived that the political situation affected the social dimension, and a few believed that the political instability affected the political dimension. All male respondents perceived that the political instability had a negative impact on women's empowerment as a whole, whereas one-third of respondents believed that the political instability affected the economic dimension of women's empowerment. A few of the men perceived that the political situation affected women's social dimension; others believed that political instability affected the political dimension negatively; and finally a few men believed that political instability had no impact on women's empowerment.

3.9 Findings of CBT Project Managers

This research has analysed examples of CBT projects, which were identified by project managers as CBT that integrates gender and focuses on women's empowerment throughout the project cycle. Project managers were permitted to define 'empowerment' as they wished, and then examined the empowerment dimensions to seek data on the kind of empowerment. Then they were asked questions that on economic, social, political and psychological empowerment achieved through their projects. These data were then analysed to identify the most frequently mentioned criteria to measure gender integration in CBT projects in Bethlehem governorate. The data were collected from the following five projects that were related to CBT: the United Nations Educational, Scientific and Cultural Organization (UNESCO) site at Ministry of Tourism and Antiquities (MOTA); Rozanna Association; Masar Ibrahim; a private project in Battir; and a rural tourism project. The five projects or organizations related to the CBT variously focus on economic development of the community, environmental ecology, heritage, culture and tourism governance, respectively. Table 3.2 summarizes the findings.

The five projects described the dimensions of empowerment used in this research. Economic empowerment is a major component in the five CBT projects where almost half of the beneficiaries are women. The projects do not include a separate budget for women's benefit (gender budget) but there are special activities targeting only women that eventually lead to their benefit and economic empowerment.

Women are generating income from the CBT projects and are gaining income independence. On the social level, women are empowered in the community and play a major role in the organizations. The projects enhance this by inviting them to network with other localities, social institutions and among other stakeholders. The psychological level of empowerment is revitalized due to the fact that women are decisive in the tourism committees inside the local authority. They are being invited to participate in the decision-making process and equality is advocated. Finally, on political empowerment, all project managers agreed on the role of women in committees and in the fact that they are significant. Another issue is that this varies from one culture to another and in private projects it is more a form of business relation where women just achieve required tasks.

Some of the women working at CBT projects were asked about their opinion on gender integration: 'Women are half the society and a mother for the other half'. Another woman stated, 'Half of the society are women and women have same right[s] and responsibilities within society, which enhances society development and developing all sectors'.

Table 3.2. Results of five CTB activities.

CBT projects	Gender integration
Three CBT projects	Gender inclusion in their proposal, while the other two CBT projects include community (male and female) as a major component
Three CBT projects	Gender is part of their vision and a necessity due to the major role of women in enriching the potential product offered to tourists
All five CBT projects	Indicates female beneficiaries to these projects, either by having a 50–50 quota or via services that women offer through this CBT project, based on women's abilities to provide these services
All five CBT projects	Emphasis on economic development, access to resources, enhance freedom, dignity and decision making
All five CBT projects	Facilitate women's empowerment through creating jobs, capacity building, identifying various roles expected from various segments of women, ensure women's efficient participation in the assessment and planning stages
Two CBT projects	Political situation adversely affects women's empowerment; other CBT projects are not affected by the political situation

Nevertheless, despite the dimensions of empowerment mentioned in these CBT projects, many CBT managers faced various difficulties in empowering women. When they were asked about the limited participation of females in the CBT projects the managers revealed the following barriers: social and religious barriers; level of capacity building; limited number of community-based initiatives; and the lack of financial resources. On the question of opportunities for effective women's participation in CBT, the project managers mentioned that these vary from simple opportunities to complex ones. They can work in homestay initiatives, local production or managerial positions at a CBT project, or as local tour guides or event managers.

Another woman stated: 'Women are the core of the society and there shouldn't be gender inequality. Now women and men have different roles in society and women's role is raising [their] children.'

A CBT male manager stated: 'A woman is not just a number but the main actor in the development of the society. CBT supports gender integration in order to empower women and achieve gender equality in all projects. The gender integration creates different perspectives and enhances decision making.'

The reason for the low level of empowerment in these three dimensions could be related to various causes such as the political situation (Israeli occupation), as indicated by men and women who emphasized that the political situation impacts their economic empowerment specifically and empowerment in general–whether social, political or psychological. Based on the findings, various CBT managers had a superficial understanding of empowerment with its all dimensions.

Another manager said:

> Gender integration is only theoretical but not included in the implementation–there is no equal opportunity. Gender integration is required for social development. Women should not be isolated from society since they are part of society and should be integrated into all projects especially the participation of youth. On the other hand, women have reached equality in employment and men now are facing problems in getting job opportunities.

In addition, the low level of empowerment could also be due to the misconceptions about and understanding of empowerment, gender integration, participatory approach and other important terms. Or it could be a trust issue related to women's and men's attitudes presented during discussion. For example, the finding indicated that stakeholders proposed initiatives to increase women's participation in the tourism sector. The types of initiatives proposed by the stakeholders clearly indicate that each group proposed initiatives based on attitudes and not on the basis of a comprehensive and well-coordinated scientific approach to the issue of gender integration through CBT. There were no common objectives that aimed to manifest sustainable development for all stakeholders and communities.

3.10 Conclusion and Implications

This study aimed to explore the factors leading to women's low participation in CBT projects in Bethlehem, identify the challenges for women's participation and highlight some initiatives that could be taken to lead to women's empowerment through tourism. The research conducted shows a positive connection between CBT and female participation. The environment surrounding CBT is conducive to women's participation in tourism.

The local economy in the city of Bethlehem has been historically associated with tourism. Family businesses were established around tourism. The Bethlehem Governorate has been and still is part of this economic structure based on tourism.

Many women from villages surrounding the city of Bethlehem came with their produce from surrounding villages to sell them to tourists. Again, craftsmanship became a family tradition, and women in Bethlehem have been part of this CBT economic dynamics. Directly or indirectly, women in the Governorate of Bethlehem got pulled into these economic dynamics. The shops need female workers to help attract tourists, and families prefer employing their own daughters as part of the family business to hiring strangers. The craftsmen need the support of their wives and daughters and so on through the value chain of the CBT economic structure. Gender integration and mainstreaming strategies on the other hand showed low and limited levels of female participation in CBT.

All five (CBT) projects in Bethlehem Governorate are driving forces to increase women's participation in tourism since the business environment allows for relative social, political, economic and psychological conditions for Bethlehem Governorate to be part of the economic dynamics surrounding CBT.

This study shows that gender integration approaches are not implemented throughout CBT development projects in the Bethlehem Governorate. Nevertheless, the study showed that if gender integration approaches are implemented through a comprehensive process and policies, starting at a national level and employing a streaming method from top-down policies and legal framing, this might improve women's participation in CBT and enhance women's empowerment and gender equality through the CBT sector. Likewise, the study showed that for women working in the CBT sector, empowerment in the four dimensions seems superficial.

The present chapter provides both academic and managerial implications. Observing the study from an academic perspective, the aim is to close the gap in the existing literature by examining women's empowerment through CBT and their participation in the tourism sector in Bethlehem. So far, to the best of the author's knowledge, no previous study has been done on this topic. Major consequences of empowerment, apart from the emancipation of women, are a change in the role of men and a perceived impact on the traditional Palestinian power structure. This study displays that the concept of empowerment is a negotiated movement, involving consultation, steady alterations and development. Based on this research, new studies must be undertaken focusing on different parts of Palestine, where there are pre-existing city or village-based institutions to promote empowerment.

This study was preliminary research that aimed to shed light on Palestinian women's participation in the tourism sector. This research could be a starting point for bringing more attention to the issue of women in tourism in the academic and political agenda. The interviews conducted in this research have shown other issues related to women in the tourism sector that could be further explored in future research. These issues include the exploitation of women, women's labour rights in CBT and informal work, women's product pricing, women's salaries compared to men's and property ownership. Empowerment and its realization began with economic empowerment, growing gradually to include social, psychological and political empowerment (Scheyvens, 1999). Strategies such as peer training must be introduced through earlier structures to initiate discussions and negotiations to promote small business development and empowerment among indigenous and female communities in Bethlehem and Palestine in general.

In the future, several factors will need to be addressed to ensure that ongoing improvements for local people result from CBT projects. Mechanisms that support the inclusion of less-educated women, the economically disadvantaged and minorities will need to be further explored. For example, in additional to hospitality training, small financial loans could be necessary including microcredit and microfinancing to facilitate the expansion of additional CBT projects within the region, to enable the inclusion of women at a lower socio-economic level during this enterprise opportunity (Kalim and Salahuddin, 2011). To be able to implement these initiatives, gender needs to be considered in tourism development and planning. As briefly mentioned in this study, some donors have demonstrated a commitment to promoting gender equality and women's empowerment through tourism. The Japan International Cooperation Agency (JICA) funds several projects that integrate gender and tourism development. Therefore, it is crucial to include gender in the tourism development and planning in Palestine. However, prior to this, there is a need for more analysis of the different (contradictory) perspectives with which gender is understood during the different stages of the development process, how the interpretations of gender can influence the success or the outcomes of a project and how we can achieve the goals of integrating indigenous women into the process of tourism development.

References

Abdo, N. (1999) Gender and politics under the Palestinian authority. *Journal of Palestine Studies* 28(2), 38–51. DOI: 10.2307/2537933.

Alarcón, D.M. and Cole, S. (2019) No sustainability for tourism without gender equality. *Journal of Sustainable Tourism* 27(7), 903–919. DOI: 10.1080/09669582.2019.1588283.

Al-Oun, S. and Al-Homoud, M. (2008) The potential for developing community-based tourism among the Bedouins in the Badia of Jordan. *Journal of Heritage Tourism* 3(1), 36–54. DOI: 10.1080/1743873X.2008.9701249.

Annes, A. and Wright, W. (2015) 'Creating a room of one's own': French farm women, agritourism and the pursuit of empowerment. *Women's Studies International Forum* 53, 1–11. DOI: 10.1016/j.wsif.2015.08.002.

Arafeh, N. (2016) *Economic Collapse in East Jerusalem: Strategies for Tecovery*. Alshabka, the Palestinian Policy Network, Washington, DC.

Arroyo, C.G., Barbieri, C., Sotomayor, S. and Knollenberg, W. (2019) Cultivating women's empowerment through agritourism: evidence from Andean communities. *Sustainability* 11(11), 3058. DOI: 10.3390/su11113058.

Bárcena-Martín, E. and Moro-Egido, A.I. (2013) Gender and poverty risk in Europe. *Feminist Economics* 19(2), 69–99. DOI: 10.1080/13545701.2013.771815.

Baumann, H. (2016) Enclaves, borders, and everyday movements: Palestinian marginal mobility in East Jerusalem. *Cities* 59, 173–182. DOI: 10.1016/j.cities.2015.10.012.

Bello, F.G., Lovelock, B. and Carr, N. (2017) Constraints of community participation in protected area-based tourism planning: the case of Malawi. *Journal of Ecotourism* 16(2), 131–151. DOI: 10.1080/14724049.2016.1251444.

Belsky, J.M. (1999) Misrepresenting communities: the politics of community-based rural ecotourism in Gales Point Manatee, Belize. *Rural Sociology* 64(4), 641–666. DOI: 10.1111/j.1549-0831.1999.tb00382.x.

Berkowitz, M. (1997) *Western Jewry and the Zionist Project 1914–1933*. Cambridge University Press, Cambridge, UK.

Bishop, D. and Bowman, K. (2014) Still learning: a critical reflection on three years of measuring women's empowerment in Oxfam. *Gender & Development* 22(2), 253–269. DOI: 10.1080/13552074.2014.920993.

Boonabaana, B. (2014) Negotiating gender and tourism work: women's lived experiences in Uganda. *Tourism and Hospitality Research* 14(1–2), 27–36. DOI: 10.1177/1467358414529578.

Boserup, M. (1970) *Woman's Role in Economic Development*. Allen and Unwin, London.

Çakmak, E. and Isaac, R.K. (2012) What destination marketers can learn from their visitors' blogs: an image analysis of Bethlehem, Palestine. *Journal of Destination Marketing & Management* 1, 124–133. DOI: 10.1016/j.jdmm.2012.09.004.

Castelberg-Koulma, M. (1991) Greek women and tourism: women's cooperatives as an alternative form of organization. In: Redclift, N. and Sinclair, M.T. (eds) *Working Women. International Perspectives on Labour and Gender Ideology*. Routledge, London, pp. 197–223.

Cave, P. and Kilic, S. (2010) The role of women in tourism employment with special reference to Antalya, Turkey. *Journal of Hospitality Marketing & Management* 19(3), 280–292. DOI: 10.1080/19368621003591400.

Chant, S. (1992) Tourism in Latin America: perspectives from Mexico and Costa Rica. In: Harrison, D. (ed.) *Tourism and the Less Developed Countries*. Belhaven, London, pp. 85–107.

Charlesworth, H. and Chinkin, C. (2000) *The Boundaries of International Law: A Feminist Analysis*. Manchester University Press, Manchester, UK.

Cherayi, S. and Jose, J.P. (2016) Empowerment and social inclusion of Muslim women: towards a new conceptual model. *Journal of Rural Studies* 45, 243–251. DOI: 10.1016/j.jrurstud.2016.04.003.

Cohen-Hattab, K. (2004) Zionism, tourism, and the battle for Palestine: tourism as a political-propaganda tool. *Israel Studies* 9(1), 61–85. DOI: 10.2979/ISR.2004.9.1.61.

Cohen-Hattab, K. and Shoval, N. (2015) *Tourism, Religion, and Pilgrimage in Jerusalem*. Routledge, Abingdon, UK. DOI: 10.4324/9781315770345.

Connell, J. and Rugendyke, B. (2008) Tourism and local people in the Asia-Pacific region. In: Connell, J. and Rugendyke, B. (eds) *Tourism at the Grassroots: Villagers and Visitors in the Asia-Pacific*. Routledge, London, pp. 1–40.

Cukier, J., Norris, J. and Wall, G. (1996) The involvement of women in the tourism industry of Bali, Indonesia. *Journal of Development Studies* 33(2), 248–270. DOI: 10.1080/00220389608422465.

Daher, R. (2000) Dismantling a community's heritage – heritage tourism: conflict, inequality, and search for social justice in the age of globalisation. In: Robinson, M., Evans, N., Long, P., Sharpley, R. and Swarbrooke, J. (eds) *Tourism and Heritage Relationships: Global, National and Local Perspectives*. University of Northumbria, Newcastle, UK, pp. 105–128.

Dahles, H. and Bras, K. (1999) Entrepreneurs in Romance tourism in Indonesia. *Annals of Tourism Research* 26(2), 267–293. DOI: 10.1016/S0160-7383(98)00098-X.

Dangi, T.B. and Jamal, T. (2016) An integrated approach to sustainable community-based tourism. *Sustainability* 8(5), 475. DOI: 10.3390/su8050475.

Dolezal, C. and Novelli, M. (2020) Power in community-based tourism: empowerment and partnership in Bali. *Journal of Sustainable Tourism* 30(10), 2352–2370. DOI: 10.1080/09669582.2020.1838527.

Duffy, L.N., Kline, C.S., Mowatt, R.A. and Chancellor, H.C. (2015) Women in tourism: shifting gender ideology in the DR. *Annals of Tourism Research* 52, 72–86. DOI: 10.1016/j.annals.2015.02.017.

Duflo, E. (2012) Women empowerment and economic development. *Journal of Economic Literature* 50(4), 1051–1079. DOI: 10.1257/jel.50.4.1051.

Ernudd, A. (2007) The role of gender in the Israeli–Palestinian conflict – exploring the impact of occupation and patriarchy on Palestinian women. Available at: https://www.lunduniversity.lu.se/lup/publication/1321164 (accessed 4 April 2019).

Ferguson, L. (2010a) Interrogating 'gender' in development policy and practice: the World Bank, tourism and microenterprise in Honduras. *International Feminist Journal of Politics* 12(1), 3–24.

Ferguson, L. (2010b) Tourism development and the restructuring of social reproduction in Central America. *Review of International Political Economy* 17(5), 860–888. DOI: 10.1080/09692290903507219.

Ferguson, L. and Alarcón, D.M. (2015) Gender and sustainable tourism: reflections on theory and practice. *Journal of Sustainable Tourism* 23(3), 401–416. DOI: 10.1080/09669582.2014.957208.

Garcia-Ramon, M.D., Canoves, G. and Valdovinos, N. (1995) Farm tourism, gender and the environment in Spain. *Annals of Tourism Research* 22(2), 267–282. DOI: 10.1016/0160-7383(94)00096-4.

Gentry, K.M. (2007) Belizean women and tourism work. Opportunity or impediment? *Annals of Tourism Research* 34(2), 477–496.

Goodwin, H. and Santilli, R. (2009) Community-based tourism: a success. *ICRT Occasional Paper* 11(1), 37.

Haj, S. (1992) Palestinian women and patriarchal relations. *Signs: Journal of Women in Culture and Soceity* 17(4), 761–778. DOI: 10.1086/494763.

Haj-Yahia, M.M. (2005) On the characteristics of patriarchal societies, gender inequality, and wife abuse: the case of Palestinian Society. *Adalah's Newsletter* 20, 1–6.

Harris, J.M., Wise, T.M., Gallagher, K.P. and Goodwin, N.R. (2001) *A Survey of Sustainable Development: Social and Economic Dimensions*. Island Press, Washington, DC.

Huete, R., Brotons, M. and Sigüenza, M.C. (2016) La desigualdad entre mujeres y hombres en el sector hostelero español. Inequality between women and men in the Spanish hospitality sector, *Estudios y Perspectivas En Turismo* 25(1), 73–87.

Ibrahim, N. (2013) *Tourism and Heritage in Palestine: A Conflict between Two Narratives*. Alternative Tourism Group, Beit Sahour, Palestine.

Isaac, R.K. (2009) Alternative tourism: can the segregation wall in Bethlehem be a tourist attraction? *Tourism and Hospitality Planning & Development* 6(3), 247–254. DOI: 10.1080/14790530903363381.

Isaac, R.K. (2010a) Alternative tourism: new forms of tourism in Bethlehem for the Palestinian tourism industry. *Current Issues in Tourism* 13(1), 21–36. DOI: 10.1080/13683500802495677.

Isaac, R.K. (2010b) Moving from pilgrimage to responsible tourism: the case of Palestine. *Current Issues in Tourism* 13(6), 579–590. DOI: 10.1080/13683500903464218.

Isaac, R.K. (2013) Palestine: tourism under occupation. In: Butler, D. and Wantanee, S. (eds) *Tourism and War*. Routledge, London, pp. 143–158.

Isaac, R.K. (2018) Religious tourism in Palestine: challenges and opportunities. In: Butler, R. and Suntikul, W. (eds) *Tourism and Religion*. Channel View Publications, Bristol, UK, pp. 143–158.

Isaac, R.K. and Ashworth, G.. (2012) Moving from Pilgrimage to dark tourism: leveraging tourism in Palestine. *Tourism Culture & Communication* 11(3), 149–164. DOI: 10.3727/109830412X1334687 6802158.

Isaac, R.K. and Eid, T.A. (2019) Tourists' destination image: an exploratory study of alternative tourism in Palestine. *Current Issues in Tourism* 22(12), 1499–1522. DOI: 10.1080/13683500.2018.1534806.

Isaac, R.K., Hall, C.M. and Higgins-Desbiolles, F. (eds) (2016) *The Politics and Power of Tourism in Palestine*. Routledge, Abingdon, UK. DOI: 10.4324/9781315740508.

Isaac, R.K. and Platenkamp, V. (2010) Ethnography of hope in extreme places: Ahrendt's Agora in controversial tourism destinations. *Tourism Culture & Communication* 12(2), 173–186. DOI: 10.3727/1098 30413X13575858951365.

Johnston, A. (2003) Self-determination: exercising indigenous rights in tourism. In: Singh, S., Timothy, D.J. and Dowling, R.K. (eds) *Tourism in Destination Communities*. CAB International, Wallingford, UK, pp. 115–134.

Joo, D., Woosnam, K.M., Strzelecka, M. and Boley, B.B. (2020) Knowledge, empowerment, and action: testing the empowerment theory in a tourism context. *Journal of Sustainable Tourism* 28(1), 69–85. DOI: 10.1080/09669582.2019.1675673.

Kalim, R. and Salahuddin, T. (2011) Micro financing of NGOs and government: collaborative impact on poverty eradication. *Information Management and Business Review* 2(2), 81–91. DOI: 10.22610/ imbr.v2i2.885.

Klasen, S. (2002) Low schooling for girls, slower growth for all? Cross-country evidence on the effect of gender inequality in education on economic development. *World Bank Economic Review* 16(3), 345–373. DOI: 10.1093/wber/lhf004.

Klasen, S. and Lamanna, F. (2009) The impact of gender inequality in education and employment on economic growth: new evidence for a panel of countries. *Feminist Economics* 15(3), 91–132. DOI: 10.1080/13545700902893106.

Knight, D.W. and Cottrell, S.P. (2016) Evaluating tourism-linked empowerment in Cuzco. *Annals of Tourism Research* 56, 32–47. DOI: 10.1016/j.annals.2015.11.007.

Kousis, M. (1989) Tourism and the family in a rural Cretan community. *Annals of Tourism Research* 16(3), 318–332. DOI: 10.1016/0160-7383(89)90047-9.

Kuttab, E. (2010) Empowerment as resistance: conceptualizing Palestinian women's empowerment. *Development Studies* 53(2), 247–253. DOI: 10.1057/dev.2010.22.

Lapeyre, R. (2010) Community-based tourism as a sustainable solution to maximise impacts locally? The Tsiseb Conservancy case, Namibia. *Development Southern Africa* 27(5), 757–772. DOI: 10.1080/0376835X.2010.522837.

Laszlo, S., Grantham, K., Oskay, E. and Zhang, T. (2020) Grappling with the challenges of measuring women's economic empowerment in intrahousehold settings. *World Development* 132, 104959. DOI: 10.1016/j.worlddev.2020.104959.

López-Guzmán, T., Borges, O. and Castillo-Canalejo, A.M. (2011) Community-based tourism in Cape Verde – a case study. *Tourism and Hospitality Management* 17(1), 35–44. DOI: 10.20867/thm.17.1.3.

Malema, D.R. and Naidoo, S. (2017) The role of community arts and crafts in the empowerment of women living in a rural environment. *World Leisure Journal* 59(suppl.1), 54–60. DOI: 10.1080/16078055.2017.1393878.

Malhotra, A., Schuler, S.R. and Boender, C. (2002) Measuring women's empowerment as a variable in international development. June. Paper prepared for the World Bank Workshop on Poverty and Gender: New Perspectives.

Maruyama, N.U. and Woosnam, K.M. (2020) Representation of 'mill girls' at a UNESCO World Heritage Site in Gunma, Japan. *Journal of Sustainable Tourism* 29(2–3), 277–294. DOI: 10.1080/09669582.2020.1738443.

Moswete, N. and Lacey, G. (2015) Women cannot lead: empowering women through cultural tourism in Botswana. *Journal of Sustainable Tourism* 23(4), 600–617. DOI: 10.1080/09669582.2014.986488.

Movono, A. and Dahles, H. (2017) Female empowerment and tourism: a focus on businesses in a Fijian village. *Asia Pacific Journal of Tourism Research* 22(6), 681–692. DOI: 10.1080/10941665.2017.1308397.

Müller, S., Huck, L. and Markova, J. (2020) Sustainable community-based tourism in Cambodia and tourists' willingness to pay. *Austrian Journal of South East Asian Studies* 13(1), 81–101.

Nabbout, M. (2017) Palestine is the world's fastest growing tourist destination. Available at: https://stepfeed.com/palestine-is-the-world-s-fastest-growing-tourist-destination-7471 (accessed 3 May 2019).

Organisation of Islamic Cooperation (2014) *Recommendations and Proposals for Supporting the Palestinian People in East Jerusalem*. SESRIC, Ankara, Turkey.

O'Hara, C. and Clement, F. (2018) Power as agency: a critical reflection on the measurement of women's empowerment in the development sector. *World Development* 106, 111–123. DOI: 10.1016/j.worlddev.2018.02.002.

PCBS (2017) Number of enterprises, employed persons and main economic indicators for the tourism enterprises in Palestine by tourism activity, 2017. Palestinian Central Bureau of Statistics. Available at: http://www.pcbs.gov.ps/Portals/_Rainbow/Documents/TourAct-2017-E-01.html (accessed 2 July 2019).

Perkins, D.D. and Zimmerman, M.A. (1995) Empowerment theory, research, and application. *American Journal of Community Psychology* 23(5), 569–579. DOI: 10.1007/BF02506982.

Prior, M. (1997) *The Bible and Colonialism: a Moral Critique*. Sheffield Academic Press, Sheffield, UK.

Rowlands, J. (1995) Empowerment examined. *Development in Practice* 5(2), 101–107. DOI: 10.1080/0961452951000157074.

Salazar, N. (2012) Community-based cultural tourism: issues, threats and opportunities. *Journal of Sustainable Tourism* 20(1), 9–22. DOI: 10.1080/09669582.2011.596279.

Scheyvens, R. (1999) Ecotourism and the empowerment of local communities. *Tourism Management* 20(2), 245–249. DOI: 10.1016/S0261-5177(98)00069-7.

Scheyvens, R. (2000) Promoting women's empowerment through involvement in ecotourism: experiences from the third world. *Journal of Sustainable Tourism* 8(3), 232–249. DOI: 10.1080/09669580008667360.

Scheyvens, R. (2002) *Tourism for Development: Empowering Communities*. Pearson Education Ltd, Harlow, UK.

Sinclair, M. (ed.) (1997) *Gender, Work and Tourism*. Routledge, London.

Smith, D.O. (2010) Hotel design in British Mandate Palestine: modernism and the zionist vision. *Journal of Israeli History* 29(1), 99–123. DOI: 10.1080/13531041003595035.

Sofield, T.H.B. (2003) *Empowerment for Sustainable Tourism Development*. Elsevier Science, Oxford, UK.

Su, M.M., Wall, G., Ma, J., Notarianni, M. and Wang, S. (2020) Empowerment of women through cultural tourism: perspectives of Hui minority embroiderers in Ningxia, China. *Journal of Sustainable Tourism*. DOI: 10.1080/09669582.2020.1841217.

Tucker, H. and Boonabaana, B. (2012) A critical analysis of tourism, gender and poverty reduction. *Journal of Sustainable Tourism* 20(3), 437–455. DOI: 10.1080/09669582.2011.622769.

UNWTO (2011) Global report on women in tourism 2010. Available at: http://ethics.unwto.org/content/global-report-women-tourism-2010 (accessed 6 November 2018).

Whitelam, K.W. (2013) *The Invention of Ancient Israel: The Silencing of Palestinian History*. Routledge, London. DOI: 10.4324/9781315812649.

4 Sustainable Tourism Development in Rural Areas: The Case of Community-based Lodges in Iran

Sarasadat Makian[1]*, Ashkan Borouj[2] and Farid Hanifezadeh[2]
[1]Grenoble Alpes University, Grenoble, France; [2]Allameh Tabataba'i University, Tehran, Iran

4.1 Introduction

Rural areas have a low population density and a predominance of agriculture as a primary sector. They differ in economic, socio-cultural and institutional terms (Straka and Tuzová, 2016). In addition, rural areas are often characterized by a lack of facilities and are inhabited by people with low incomes (Giampiccoli and Kalis, 2012b). Tourism contributes to the development of rural areas, diversifies the rural economy and contributes to sustainable rural development (Bel *et al.*, 2015; Nooripoor *et al.*, 2020). Community-based tourism (CBT), in particular, is emerging as a vehicle for sustainable rural development in remote areas (Zielinski *et al.*, 2020). CBT's revenues are vital for local people (Giampiccoli and Kalis, 2012b). The large majority of CBT initiatives focus on developing community-owned and -managed lodges (Goodwin and Santilli, 2009). They have significant potential for generating income, empowerment and self-sufficiency (Mtapuri and Giampiccoli, 2013). A community-based lodge (CBL) development needs to balance three key elements of sustainable tourism, namely economic, socio-cultural and environmental sustainability (Mearns, 2011).

In developing countries, tourism is an activity that brings necessary economic benefits to communities and encourages local people to participate in the development process (Yanes *et al.*, 2019). Iran is one of those developing countries that needs to revitalize its economy through tourism development (Nematpour *et al.*, 2020). Although rural areas of Iran have great tourism potential, previous studies show that tourism development in these areas faces significant economic, environmental and socio-cultural sustainability issues (Kalantari *et al.*, 2008; Eftekhari *et al.*, 2011a; Dadvar-Khani, 2012; Ghaderi and Henderson, 2012). CBT often leads to various financial/non-financial benefits for local communities and maximizes sustainability at the local level (Han *et al.*, 2019). As CBT develops in rural areas of Iran, CBLs offer an excellent opportunity to attract tourists (Bayat and Khazaiepool, 2017). Several studies have been published assessing the sustainability of tourism in rural areas of Iran (Eftekhari *et al.*, 2011b; Yarihesar and Bakhtar, 2016; Taleshi and Khodapanah, 2020), but none has determined the sustainability indicators for CBL development. This study aims to identify the sustainability indicators for CBL development. The development of

*Corresponding author: sarasadat.makian@univ-grenoble-alpes.fr

 Tourism Planning and Development in the Middle East (eds S. Kladou *et al.*)
DOI: 10.1079/9781789249149.0004

CBLs enables the preservation and revival of rural cultural heritage (tangible and intangible) and the natural environment. The main purpose of creating and developing this type of accommodation is to achieve sustainable tourism development (Radioeqtesad, 2018). The Iranian Ministry of Cultural Heritage, Tourism and Handicrafts also prioritizes small businesses in rural areas and similar initiatives to develop the Iranian tourism and hospitality industry (*Tehran Times*, 2017).

This chapter focuses on concepts such as community-based rural tourism and CBLs. We highlight the development of CBT in Iran. Then we provide an overview of these types of accommodation in rural areas of Iran. Using the Delphi and Analytic Hierarchical Process (AHP) methods, we examine how adequate development of CBLs can contribute to the sustainable development of rural areas. Accordingly, we identify sustainability indicators for CBL development in rural areas.

4.2 Sustainable Tourism Development in Rural Areas

Emigration and depopulation are common problems in peripheral rural areas (Stockdale, 2004; Adamiak *et al.*, 2017). Tourism development can help reverse this trend by contributing to these areas' local economic and socio-cultural revitalization and improving residents' quality of life (Eusebio *et al.*, 2014). This development also reduces poverty, immigration and allows the preservation of valuable traditions (Yousefi, 2016). The contribution of tourism to rural development lies in its strong link with various rural resources and its integration into local economies to achieve many development benefits. Therefore, rural areas have become a crucial component of tourism and recreational development (Aslam and Awang, 2015). Consequently, tourism in rural areas has attracted growing interest from academics, tourism professionals, investors and politicians for many reasons; due to the existing market forces and government initiatives (Lane, 1994), it is seen as a potential development tool for often economically weak areas (Eusebio *et al.*, 2014). Rural tourism highlights rural life, art, culture and heritage. It supports the local community economically and socially. It also allows interaction between tourists and locals to enrich the tourist experience (Sasu and Epuran, 2016). The development of tourism in rural areas diversifies the rural economy, provides the necessary framework for sustainable rural development and stimulates the national economy's growth (Mirani and Farahani, 2015). In addition, rural tourism development can satisfy both supply and demand parties without compromising natural, cultural and social resources (Eusebio *et al.*, 2014).

Since tourism in rural areas is highly dependent on environmental attractiveness and healthy outdoor activities, one would expect that tourism and hospitality operators would be motivated to engage in sustainable development practices (Carlsen *et al.*, 2001). Sustainability principles concern the appropriate balance between environmental, economic and socio-cultural aspects of tourism development (Korez-Vide, 2013). Sustainable development of tourism prevents the deterioration of the rural environment's physical, social and economic life by conserving natural resources and creating a recreational tourist space with minimal manipulation of the natural environment (Yousefi, 2016). In rural areas, sustainable tourism is often achieved through small-scale tourism-centred localized projects that focus on the sustainable use of resources (Sharpley, 2007). This aims to present a new strategy to revitalize or transform rural areas based on natural and human resources through complementary activities (Shamsoddini, 2015). One of the key elements of tourism development in these areas is to inspire greater involvement of the local community as a basis for the destination's sustainability (Juma and Khademi-Vidra, 2019).

4.3 Community-based Rural Tourism

A tourism-based approach to community development contributes to rural areas' development (Sebele, 2010; Keyim, 2017). CBT promotes local development in developing countries, particularly in rural areas (Davison *et al.*, 2005). In order to support the development of CBT, it is essential to involve local community members (Thetsane, 2019) and encourage community identity (Polnyotee and Thadaniti, 2015). In this way, community quality of life improves (Yu *et al.*, 2016), and environmental sustainability gets better by enhancing the preservation of natural resources (Sutresna *et al.*, 2019). CBT is related to the uniqueness of local culture (Giampiccoli and Kalis, 2012a) and the development of cross-cultural learning and cultural differences, as well as human dignity (Purbasari and Manaf, 2018). CBT development should distribute economic, social and environmental benefits equally. Economic benefits help generate wealth for local communities through income (Yanes *et al.*, 2019), leading many rural communities to turn to tourism to diversify their economic base (Johnson, 2010).

CBT projects are collaborative efforts of local communities to develop small- to medium-scale tourism. This development is compatible with the rural context by taking into account environmental, social and cultural sustainability (Johnson, 2010; Yanes *et al.*, 2019). The locals run and own CBT activities to benefit the community, allowing tourists to become aware of and informed about the community and local lifestyles (Fazel Bakhsheshi and Hassanali, 2018). Community-based rural tourism (CBRT) is an extension of CBT that occurs in a rural setting (Kayat, 2014) and is one of the components of an overarching plan to improve rural economies (Johnson, 2010). Typically, within the framework of CBT, the community manages the activities a tourist engages in, such as accommodation, food, guidance and the sale of handicrafts (Giampiccoli and Kalis, 2012b). CBRT needs the community's participation in managing tourism activities within the community, which ultimately contributes to its well-being and indirectly promotes its culture (Kayat, 2014). Although non-local intermediaries may sometimes manipulate the local context for marketing or tourists' perceived tastes, CBRT allows locals to express and represent their own culture by organizing and managing tourism activities (Giampiccoli and Kalis, 2012b).

4.3.1 Community-based rural tourism in Iran

Rural tourism, as a development strategy, is relatively new in Iran. It dates from the beginning of the 21st century (Ghaderi and Henderson, 2012). Despite the great potential of Iran's rural areas for tourism development (Ghaderi and Henderson, 2012; Akbarian Ronizi *et al.*, 2016) and the economic and social importance of tourism for this country, its development has been poorly managed (Roshan *et al.*, 2018). The country economically depends on oil; thus, tourism development can help Iran shift away from a single-product economy (Eftekhari and Ghaderi, 2002; Akbarian Ronizi *et al.*, 2016; Roshan *et al.*, 2018). Despite the positive economic benefits and the significant role of rural tourism in reversing rural–urban migration (*Financial Tribune*, 2017), the Iranian government has paid little attention to rural tourism in rural development planning and policy making (Bayat *et al.*, 2019).

Clear policy leadership and commitment to tourism are of the main pillars of development in the Middle East region (Kovjanic, 2014), which possesses excellent diversity and heterogeneity in tourism (Morakabati, 2013). As such, this study focuses on Iran, one of the most promising tourist destinations in the Middle East. Iran has unspoiled nature, diverse, picturesque landscapes, a 1000-year-old cultural and architectural heritage, a variety of recreational opportunities as well as an authentic character, rural charm and natural beauty, especially in intact rural areas (Aref and Gill, 2009; Eghbali *et al.*, 2011; Pakshir and Nasir, 2011). Living culture is also primarily found in Iran's rural areas, where art, crafts, costumes, cuisine, music and traditions of diverse ethnic communities are less affected by tourism development (O'Gorman *et al.*, 2007).

Previous studies indicate that rural tourism and CBT are the most appropriate tourism development types because of Iran's national, social, cultural, political and economic conditions (Asadi and Daryaie, 2012; Behzadfar *et al.*, 2018). While there are challenges in securing rural people's livelihoods in Iran, CBT is an important way to overcome poverty and increase participation in community activities

(Hesam and Baghiani, 2018). As the unemployment rate in Iran's rural communities increases, CBT's development can reduce this rate and distribute tourism revenue among the population. In addition, the development of CBT creates new tourism products based on rural people's lifestyles, cultures and customs. This allows rural people to do their best in preserving and presenting these products to tourists (Behzadfar *et al.*, 2018).

4.4 CBLs as a Small Family Tourism Business

Tourism contributes to sustainability by empowering local communities. This can occur by involving the community in decision making and enabling them to determine developmental objectives (Graci, 2012). Tourism also offers opportunities for many family businesses, often incorporating direct interaction between hosts and guests in the family home or property. These interactions are essential for tourists' experience and satisfaction and destination and community development (Getz and Carlsen, 2005). CBT concerns many individually or collectively managed small and micro-enterprises at the local level. Therefore, the expectation is to benefit the entire community through tourism businesses, including those indirectly involved, due to improving and preserving local standards and heritage, respectively, as well as the environment (Mtapuri and Giampiccoli, 2013). However, these family and owner-operated businesses face many difficulties in remaining commercially viable and developing their business while applying sustainable development principles (Carlsen *et al.*, 2001).

One of the CBT's main objectives is developing a new model of community-based enterprises, known as a 'community lodge' (Novelli and Gebhardt, 2007), which is a community-owned and managed housing or hostel (Goodwin and Santilli, 2009; Zielinski *et al.*, 2020). There is a clear link between the community lodges and the community, as the community benefits directly from these lodges' various products and services (Mtapuri and Giampiccoli, 2013). Local development and promotion of local small businesses would allow CBLs to contribute to overall sustainable tourism development (*Tehran Times*, 2017). This could come about by raising awareness of the potential of CBLs for the sustainable development of rural areas among local communities, stakeholders and government officials (Cheruon *et al.*, 2015).

4.4.1 CBL network in Iran

One of the efforts to develop CBRT in Iran is accommodation located in rural areas, operating as CBLs (Eftekhari and Ghaderi, 2002). The history of CBLs in Iran dates back to 2001 and 2002 (Anabestani *et al.*, 2018). Iran's CBL network consists of Iranian local tourism activists who provide accommodation and related services and create indigenous life experiences. These CBT activities improve villagers' living conditions, offering them a complementary role alongside agricultural activities (IRNA, 2016). CBLs in Iran are those accommodations established in places with unique local cultures and natural surroundings. These lodges follow the area's characteristics, such as its historical context, architectural style and natural landscape, while interaction with the local community and tourists is essential (MCTH, 2021). CBL development is an excellent opportunity for small family-based businesses in Iran to be sustainable. Direct and mutual relations take place between the guest and host family in these lodges. The presence and participation of family members are essential to satisfy the tourist experiences of rural areas (Borouj, 2016). In addition, the host family members (i.e. owners) are the ones who manage and run the lodge (Rezvani and Azizi, 2019).

CBLs offer great potential to attract tourists by connecting rural life and cultural, historical, ethical and geographical roots with natural attractions (Bayat and Khazaiepool, 2017). Developing these accommodations in Iran maintains and revives traditional Iranian architecture (Farajirad and Ehsani, 2011). The lodge's physical layout is shaped by an authentic architectural style with local materials, interior designs and local furnishings. A lodge can be a tourist attraction in its own right (Borouj, 2016). These local accommodations are mainly in old renovated houses and areas with historic sites. Each lodge preserves its own local identity. The guests try to familiarize themselves with this particular place's traditional way of life, have local food and get to know the lodge's natural environment (Rezvani and Azizi, 2019).

CBLs are of interest to the Iranian government and tourism authorities as they can provide opportunities for creating jobs for locals, reviving rural areas and developing the local economy (*Financial Tribune*, 2016). The establishment of lodges represents a major step in developing certain peripheral areas in Iran with previously low employment opportunities. For instance, more than half of the registered CBLs in Iran (58%) are in provinces with hot and dry climates and desert landscapes (MCTH, 2020). The various benefits of these lodges have contributed to generating income in disadvantaged areas and empowering women (IANA, 2017). The owners and managers of lodges are local, as are their products and architectural style, offering tourists local-based services. Tourists can choose to stay in a lodge with a nomadic or rural family or enjoy staying in independent accommodation (*Tehran Times*, 2020). Lodges do not simply have a residential function. CBLs have food and beverages, handicraft production and sale, traditional ceremonies and music performances, local events and eco-tours (Borouj, 2016).

The number of these lodges is increasing in different parts of the country. Foreign and domestic tourists favour them. CBLs are mainly aimed at cultural tourists with low to medium income levels. According to the Iranian Ministry of Cultural Heritage, Tourism and Handicrafts, 1947 CBLs were registered by 2020 (MCTH, 2020) meaning that they have almost quadrupled in number since 2013 (*Tehran Times*, 2020). In Iran, the best-known CBLs are in Fars, Hormozgan, Khuzestan, Kerman, Isfahan, Semnan, Mazandaran, Gilan and Golestan provinces (Anabestani *et al.*, 2018). Figure 4.1 presents the distribution of CBLs in Iran. The map's shading spectrum relates to the number of lodges in each province. The city of Kerman, located in the southeast of Iran, with 314 lodges, has the highest density of this type of accommodation. The city of Qom has the lowest number, with only one lodge.

In Iran, there are several examples of such accommodation. Here, we examine three

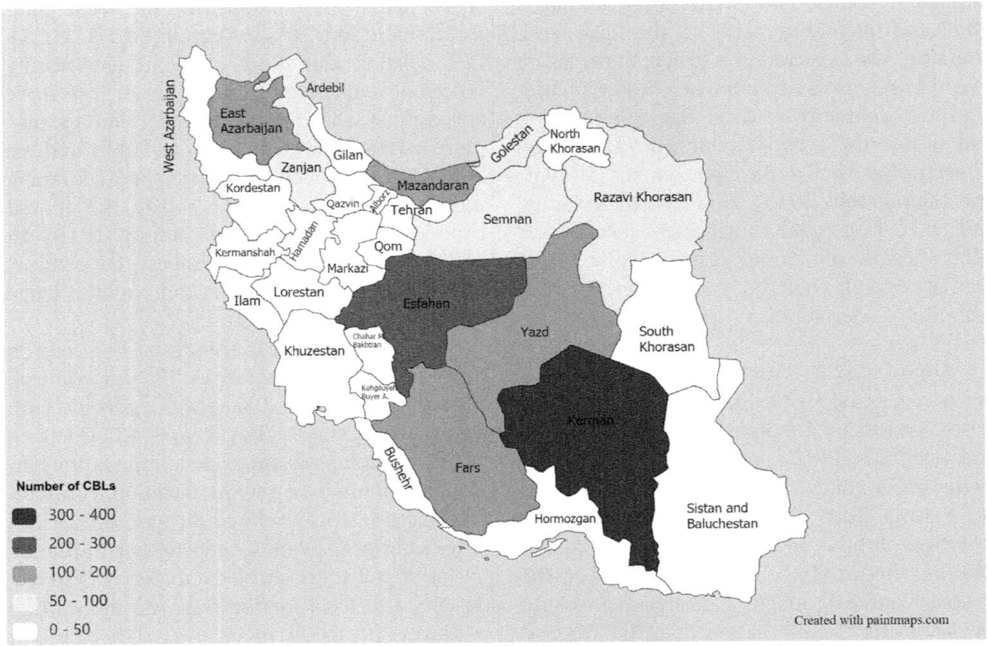

Fig. 4.1. Map of community-based lodge distribution in Iran. Created using data from MCTH, 2020.

Fig. 4.2. Lodge courtyard. Used with permission from Ashkan Borouj.

Fig. 4.3. Local Iranian food served in the lodge. Used with permission from Ashkan Borouj.

Iranian CBLs. 'Nartitee' (http://nartitee.ir/, accessed 23 October 2022) is a CBL with a traditional environment as well as a warm and intimate atmosphere, located in the centre of Iran in Yazd province (one of the Iranian sites included in the United Nations Educational, Scientific and Cultural Organization (UNESCO) World Heritage List). The young couple who run this lodge were not satisfied with their daily life, and sought new experiences and challenges by turning their old, ancestral home into a CBL. 'Ateshooni' (http://ateshooni.com/, accessed 23 October 2022) is a 300-year-old house converted into a CBL located in the central desert of Iran, in Isfahan province. This lodge offers interaction with other guests, night camping and a 4×4 safari. 'Noghli' lodge (https://noghlihouse.com/, accessed 23 October 2022) is a 150-year-old Qajar-era house in Kashan, Isfahan province. Although small, the Noghli offers its residents all the characteristics of traditional Iranian houses while having a relaxing atmosphere and offering local handicrafts. Figs 4.2–4.5 are photos of some of these lodges.

Fig. 4.4. Lodge courtyard with local furniture. Used with permission from Ashkan Borouj.

4.4.2 Iran's situation regarding COVID-19 and CBLs

The start of the COVID-19 pandemic in December 2019 had a negative impact on a variety of economic activities, human well-being and human lives throughout the world. Meanwhile, this crisis has hit the tourism industry hardest (Sigala, 2020). Iran has been the most affected in the Middle East, forcing it to impose a lockdown (Mirzaei *et al.*, 2021), resulting in a 96% decrease in international tourist arrivals during the same period (Masaeli *et al.*, 2021). Tourism SMEs have been among the worst affected sectors in Iran during this period (Haqbin *et al.*, 2022), including CBLs. They have faced layoffs, decreased occupancy rates, declining revenues, career changes to agriculture or animal husbandry and even the closure of lodgings. As a result, they had to make quick and innovative decisions to survive. Certain CBLs adopted strategies such as temporary business closures, government loans and a shift from international to domestic markets (*Financial Tribune*, 2020).

Fig. 4.5. Lodge central courtyard. Used with permission from Ashkan Borouj.

4.5 Methodology

There is a great need for assessment indicators to ensure the sustainable operation of CBLs

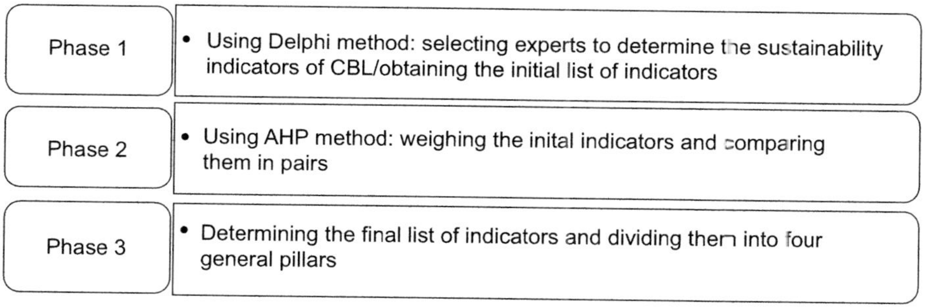

Fig. 4.6. The phases of the methodology.

due to their increasing development in Iran. Therefore, this study aims to determine the indicators to consider when setting up and managing a CBL to achieve sustainability principles in rural areas. On the other hand, as there are no studies on the sustainability of CBL development in Iran's rural areas, we decided to rely on experts' opinions for this study. A combination of Delphi and AHP was used. Delphi is an appropriate method where there are no standard criteria for assessment, and AHP is a decision-making method that prioritizes the hierarchy and consistency of judgemental data provided by a group of experts (Taleai and Mansourian, 2008). This study identified the main sustainability indicators for the development of CBLs and defined their significance through the Delphi and AHP methods, respectively, conducted in three phases (Fig. 4.6).

4.5.1 Phase 1

Through purposive sampling, 20 experts were selected based on their affiliation, years of experience and areas of expertise. In this sampling method, the researcher assigns cases to be included in the sample (Goodarzi *et al.*, 2018). The expert panel included lodge owners, eco-tour operators and university professors in the tourism and ecotourism fields. We asked experts to determine the indicators that a CBL needs in order to be sustainable, taking into account the Eftekhari *et al.* (2011a) framework for sustainable tourism development in rural areas. A study by Eftekhari *et al.* (2011a) provided a framework for evaluating tourism's sustainability with key dimensions and sustainability indicators focused on the host community (see Table 4.1).

After several rounds of discussion, based on expert opinions and agreement, we have created a primary list of sustainability indicators for developing a CBL.

4.5.2 Phase 2

Considering the AHP pair-wise comparison matrix, a self-administered questionnaire was used in the second step. The AHP expresses the relative importance of criteria through pair-wise comparisons and converts the values of pair-wise comparisons to enable them to be prioritized (Zhang *et al.*, 2014). Seven out of the previous 20 experts were selected by convenience sampling. They were individually requested to give weight to the indicators identified during the first Delphi session. The indicators with the highest scores were selected. In pairs, all indicators were compared and respondents determined which indicators were more significant. Thus, a new list of indicators has been obtained. Experts were then asked to indicate the degree of importance of each indicator according to the cells of the comparison matrices, which have numeric values from one to nine (from equally important to extremely important), in order to reflect the relative preference in each of the pairs being compared (Mu and Pereyra-Rojas, 2017).

Table 4.1. Dimensions and indicators for measuring the sustainability of tourism focusing on the host community (Eftekhari *et al*., 2011a, p. 19).

Dimensions	Sub-dimensions	Combined indicators	Indicators of tourism sustainability
Socio-cultural	Socio-cultural sustainability	Respecting human rights	Respect for guests, women and the elderly; labour rights and quality satisfaction
		Improving life quality	Development of social services and local infrastructure; quality of employment and satisfaction with income; health and hygiene; satisfaction of local community with tourism; social security
		Preserving socio-cultural patterns	Support and promote socio-cultural patterns; protect cultural heritage and historical context of the village
	Institutional and political sustainability	Local participation and monitoring	Participation in planning, decision making and implementation
		Institutional capacity	Access to structural and institutional information and communication
Economic	Economical sustainability	Equity and economic welfare	Employment; income distribution; job opportunities; poverty reduction
		Economic stability	Strengthening the local economy and long-term vitality; diversifying rural economic activities; reducing vulnerability
	Sustainability of production structure and tourism services	Tourism products and services quality	Tourist satisfaction; quality of service and tourism experience; quality of tourism products
Environmental	Natural resource sustainability	Land resources	Optimal land use, water resources and vegetation; quality of landscape
		Protecting animal and plant species	Preserving the diversity of fauna and flora
	Environmental effects	Carrying capacity	Environmental and social pressure
		Environmental vulnerability	Natural and climatic risks; preservation of heritage and natural attractions
	Environmental management and policies	Planning and management	Planning and management
		Environmental health	Waste management
		Environmental awareness	Raising environmental awareness
		Challenges at the worldwide level	Energy consumption, climate change and global warming

4.5.3 Phase 3

We reached the final list of the main sustainability indicators of the CBL after two phases of soliciting expert opinions. The indicators were then divided into four categories (pillars). Combining the obtained indicator scores from the pair-wise comparison matrices led to the definition of each pillar's significance level for the sustainable management and development of CBLs in rural areas.

4.6 Findings

Identifying 23 indicators and categorizing them into four pillars based on the Delphi sessions and AHP analysis resulted in determining the sustainability indicators for developing CBLs in rural areas. According to the experts, the core pillars of a sustainable CBL are 'indigenous tourism products, services and activities', 'ecological and environmental structure', 'family ownership and management and local community participation' and 'tourism infrastructure', as shown in Fig. 4.7. This categorization included socio-cultural, economic and environmental aspects of sustainable development. The 'tourism infrastructure' pillar is not limited to CBL development. The first pillar comprises eight indicators based on local tourism products (e.g. handicrafts), services (e.g. eco-tours) and activities (e.g. participating in local events). The second pillar, with six indicators, concerns the natural and environmental aspects of the lodge, such as the use of local materials, local internal

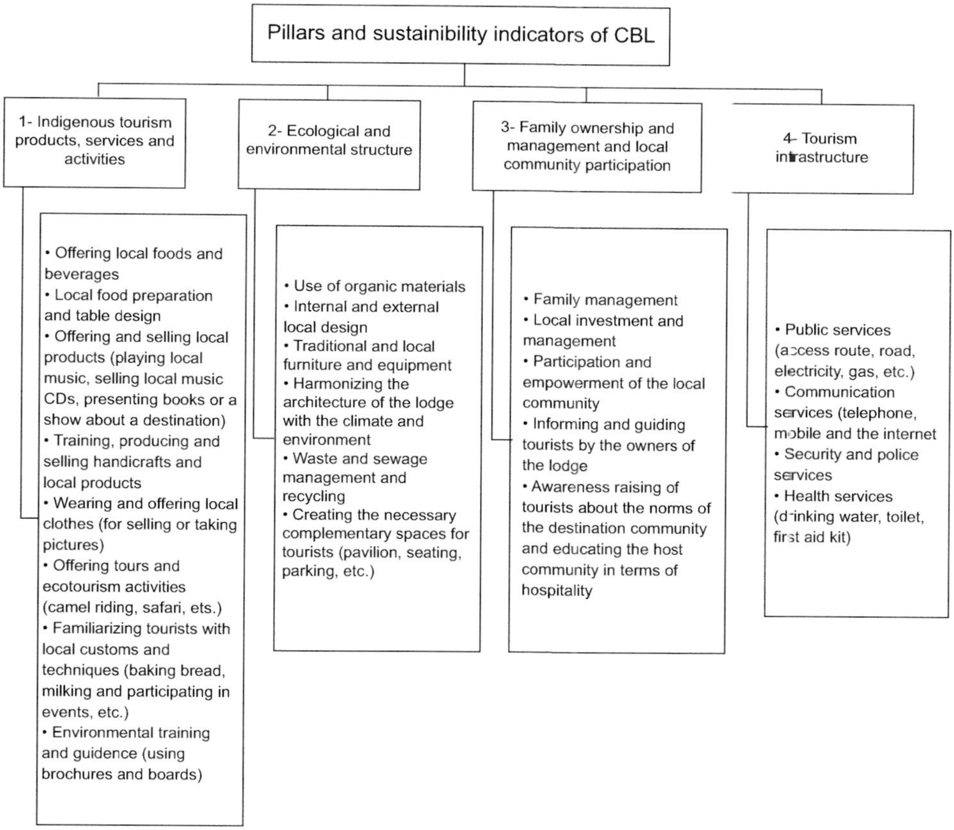

Fig. 4.7. Sustainability indicators for a community-based lodge (Developed by the authors based on research findings.).

Table 4.2. Level of significance of the core pillars of a CBL.

Core pillars	Weights
Ecological and environmental structure	0.32
Indigenous tourism products, services and activities	0.27
Family ownership, management and local community participation	0.24
Tourism infrastructure	0.17

and external design, and traditional architecture. Indicators such as family management, local investment and local people's participation are included in the third pillar with five indicators. The fourth pillar includes four tourism infrastructure indicators, such as availability of drinking water and ease of access to roads.

Table 4.2 shows the level of significance of the core pillars of a CBL based on AHP analysis. Findings showed that the 'Ecological and environmental' indicators were the most important to consider when developing and managing a CBL. The pillars of 'Indigenous tourism products, services and activities' and 'Family ownership, management and local community participation' are of second and third importance. According to experts, the 'Tourism infrastructure' pillar had the smallest significance in the sustainability of these types of lodges.

4.7 Discussion and Conclusion

The environmental, economic and socio-cultural aspects of tourism development are all part of the continuous process of achieving sustainable tourism development (Krstj *et al.*, 2008). CBT is an essential aspect of sustainable tourism development (Han *et al.*, 2019). Previous studies aimed at identifying sustainable tourism development indicators at the community level (Choi and Sirakaya, 2006; Bulatovic and Rajovic, 2016). There are also several studies on sustainability indicators of rural tourism in Iran (Eftekhari *et al.*, 2011a, b; Mahdavi *et al.*, 2016); yet, in this study, the focus was on CBL. Developing new tourism products based on rural lifestyles, culture and customs can lead to the development of CBLs as a part of a CBT project. Careful creation, management and control of CBLs are the prerequisites for sustainable tourism. The present study determined the required sustainability indicators when developing and managing a CBL in Iran's rural areas. This research attempted to propose a set of CBL development indicators to be more consistent with sustainable development principles. Based on the findings, there is a need to establish a suitable balance between the sustainability indicators of CBLs to ensure their long-term sustainability. In addition, by considering these indicators as an evaluation framework, it is possible to monitor CBL sustainability effectively. These indicators are important management tools that provide real-time information to address pressing issues and guide the destination's sustainable development (Mearns, 2015). On the other hand, the core pillars and indicators presented in Fig. 4.7, which are consistent with the pillars of sustainable development, serve as the basis for a CBL development in rural areas, particularly in Iran, as a case study of this research.

The 'Indigenous tourism products, services and activities' pillar includes indicators that promote sustainable development in Iran's rural areas by supporting socio-cultural patterns and safeguarding their intangible and cultural heritage (Wang *et al.*, 2018). The 'Ecological and environmental structure' pillar aims at preserving the historical context and cultural identity of rural areas while contributing to sustainable development (Rajković and Bojović, 2016; Taheri and Emamgholi, 2017). In terms of sustainable development, 'Family ownership and management and local community participation' indicators aim to empower local communities by involving them in planning and management, a critical issue in CBT (Khalid *et al.*, 2019). This pillar contains indicators related to community employment opportunities, strengthening the local economy and diversifying economic activities through community participation (Aulya *et al.*, 2016). The fourth pillar is concerned with tourism infrastructure. The indicators of this pillar affect tourist satisfaction (Blazeska *et al.*,

2018). Tourism infrastructure is an integral part of any development (Seetanah *et al.*, 2011). However, tourism infrastructure is only one of the factors contributing to sustainable development. Economic, socio-cultural and environmental factors all play a role (Mamirkulova *et al.*, 2020).

CBT has the potential to overcome some of the many challenges that rural and peri-urban areas face. Examples include creating employment opportunities for the local population or diversifying the existing rural economy (Bayat and Khazaiepool, 2017). If this type of tourism is not properly regulated, it jeopardizes the quality of the physical environment, social structures and cultures (Cawley and Gillmor, 2008). CBLs must take the same approach as CBT in environmental, social and cultural sustainability (Goodwin and Santilli, 2009). The development of CBLs in Iran's rural areas leads to important outcomes, such as improving the quality of the environment by preserving, developing and enhancing the local environment and culture; improving the quality of local communities' lives by involving and benefiting them; and increasing the quality of tourist experiences by creating respect, cooperation and satisfaction. All of these outcomes contribute to the growth of small tourism businesses. The proper development of these lodges supports CBT's sustainable development. In addition, CBLs, as a small tourism business with local ownership and family management, are more in line with sustainable tourism development in rural and pristine natural areas. Therefore, in this study, we identified the sustainability indicators that must be considered when establishing, managing, monitoring and evaluating a CBL. Many developing countries consider CBT to reduce poverty and build capacity at the local level (Yamashita, 2011).

On the theoretical side, as there are few studies on CBL in Iran, this study contributes to improving knowledge and filling gaps in this field, especially regarding the importance of this accommodation in the development of CBT in rural areas, as well as its role as a small and family-owned business in the sustainable development of Iranian tourism. In terms of practical implications, the findings of this study can be used to assess the level of sustainability of lodges. On the other hand, by considering the sustainability indicators in the establishment and management of CBLs, these lodges can contribute to the development of sustainable tourism. Given the rapid development of these lodges in Iran, it is recommended that the rural and community tourism authorities apply the CBL sustainability indicators to evaluate and monitor their activities and performance. In order to achieve sustainability, lodge owners, managers and local communities need to follow sustainability indicators. The indicators presented can be applied to strengthen CBT in rural areas of any destination sustainably. This type of development requires the feasibility of establishing CBLs and the willingness of the local population to take part. Furthermore, these sustainability indicators can vary depending on the cultural characteristics of rural areas. In future research it would be worth investigating other types of lodges in rural areas, such as eco-camps. There is also a need to develop an assessment framework based on the indicators identified in this study to measure the sustainability of CBLs. In this way, lodge owners and local tourism managers can ensure the sustainability of CBL in rural areas. CBLs play an important role in developing tourism in Iran's rural areas by providing indigenous tourism products, services and activities while taking ecological and environmental factors into account and allowing local communities to participate. Given that the hospitality sector is a vital part of Iran's tourism structure, the findings of this study can be used to develop recovery strategies for CBLs facing post-COVID-19 challenges (Bagheri and Abdi, 2021).

References

Adamiak, C., Pitkänen, K. and Lehtonen, O. (2017) Seasonal residence and counterurbanization: the role of second homes in population redistribution in Finland. *GeoJournal* 82(5), 1035–1050. DOI: 10.1007/s10708-016-9727-x.

Akbarian Ronizi, S.R., Roshan, G.R. and Negahban, S. (2016) Assessment of tourism climate opportunities and threats for villages located in the Northern coasts of Iran. *International Journal of Environmental Research* 10(4), 601–612.

Anabestani, A., Giyahi, H. and Javanshiri, M. (2018) An analysis of the effects of establishing ecotourism residence on development of rural settlements: a case study of Radakan village, Chenaran County. *Spatial Planning* 8(2), 1–24.

Aref, F. and Gill, S.G. (2009) Rural tourism development through rural cooperatives. *Nature and Science* 7(10), 68–73.

Asadi, R. and Daryaie, M. (2012) Strategic planning of rural tourism in Iran. *Journal of Basic and Applied Scientific Research* 2(11), 11679–11689.

Aslam, M. and Awang, K. (2015) Enterprising rural tourism for sustainable rural development in Sri Lanka. *International Journal of Economics and Financial Issues* 5(1S), 27–33.

Aulya, R., Mindarti, L.I. and Amin, F. (2016) Strengthening local economy through community-based tourism in governance perspective (study at Sidomulyo tourism village, Batu City). *Jurnal Ad'ministrare* 3(2), 11. DOI: 10.26858/ja.v3i2.2424.

Bagheri, F. and Abdi, N. (2021) Discovering recovery strategies from COVID-19 pandemic in lodging sector: insights from ecolodges and luxury hotels (the case of Guilan, Iran). *International Journal of Spa and Wellness* 4(2–3), 173–192. DOI: 10.1080/24721735.2021.1986969.

Bayat, S. and Khazaiepool, M. (2017) Values of perceptions of tourists in Mesr desert village as community-based tourism. *Economy and Rural Development* 5(18), 65–81.

Bayat, N., Rastegar, E., Salavati, L., Darabi, H., Ahmadi Fard, N. *et al.* (2019) Motivation-based market segmentation in rural tourism: the case of Sámán, Iran. *Almatourism – Journal of Tourism, Culture and Territorial Development* 10(19), 1–24.

Behzadfar, M., Akbarpour, A. and Golmakani, M. (2018) Investigating the impact of community-based tourism on rural business development (case study: Nayband village). In: *National Conference on New Patterns in Management and Business*, Jihad University Development Studies Research Institute, pp. 1–14.

Bel, F., Lacroix, A., Lyser, S., Rambonilaza, T. and Turpin, N. (2015) Domestic demand for tourism in rural areas: insights from summer stays in three French regions. *Tourism Management* 46, 562–570. DOI: 10.1016/j.tourman.2014.07.020.

Blazeska, D., Strezofski, Z. and Klimoska, A.M. (2018) The influence of tourist infrastructure on the tourist satisfaction in Ohrid. *UTMS Journal of Economics* 9(1), 85–93.

Borouj, B. (2016) Community-based tourism, a new story for nostalgic tourism. Available at: https://www.magiran.com/article/3333314 (accessed 21 August 2020).

Bulatovic, J. and Rajovic, G. (2016) Applying sustainable tourism indicators to community-based ecotourism tourist village Eco-katun Štavna. *European Journal of Economic Studies* 16(2), 309–330. DOI: 10.13187/es.2016.16.309.

Carlsen, J., Getz, D. and Ali-Knight, J. (2001) The environmental attitudes and practices of family businesses in the rural tourism and hospitality sectors. *Journal of Sustainable Tourism* 9(4), 281–297. DOI: 10.1080/09669580108667403.

Cawley, M. and Gillmor, D.A. (2008) Integrated rural tourism: concepts and practice. *Annals of Tourism Research* 35(2), 316–337.

Cheruon, R., Burugu, R. and Bor, T. (2015) Eco-lodges, a future for sustainable tourism in Kenya. *Journal of Tourism, Hospitality and Sports* 8, 38–41.

Choi, H.C. and Sirakaya, E. (2006) Sustainability indicators for managing community tourism. *Tourism Management* 27(6), 1274–1289. DOI: 10.1016/j.tourman.2005.05.018.

Dadvar-Khani, F. (2012) Participation of rural community and tourism development in Iran. *Community Development* 43(2), 259–277. DOI: 10.1080/15575330.2011.604423.

Davison, R., Harris, R.W. and Vogel, D.R. (2005) E-commerce for community-based tourism in developing countries. In: *The 9th Pacific Asia Conference on Information Systems*, pp. 1396–1403.

Eftekhari, A.R. and Ghaderi, E. (2002) The rule of rural tourism in rural development (theoretical analysis). *Journal of Spatial Planning* 6(2), 23–41.

Eftekhari, A.R., Mahdavi, D. and Pourtaheri, M. (2011a) Sustainability assessment of tourism in cultural-historical villages of Iran with an emphasis on sustainable tourism development paradigm. *Quarterly Journal of Tourism Studies* 5(14), 1–39.

Eftekhari, A.R., Pourtaheri, M. and Mahdavi, D. (2011b) Localization process of sustainable development indicators of rural tourism in Iran. *Journal of Rural Research* 1(4), 1–41.

Eghbali, N., Bakhshandekh, A. and Alipour, S.K. (2011) Investigation challenges and development of rural tourism: case study of rural Semnan province, Iran. *South Asian Journal of Tourism and Heritage* 4(1), 44–55.

Eusebio, C., Kastenholz, E. and Breda, Z. (2014) Tourism and sustainable development of rural destinations: a stakeholders' view. *Revista Portuguesa de Estudos Regionais* 36, 13–21.

Farajirad, A. and Ehsani, A. (2011) Study of Indigenous cluster experiences to make a model for expanding local network in Gramer and Shibderaz villages. *Geographical Journal of Territory* 8(30), 63–78.

Fazel Bakhsheshi, F. and Hassanali, N. (2018) Participatory tourism development in Iran: implementing community-based tourism within a migrating nomadic tribe. In: Seyfi, S. and Hall, C.M. (eds) *Tourism in Iran Challenges, Development and Issues*. Routledge, London, pp. 193–206.

Financial Tribune (2016) Iranian ecolodges increase tenfold. Available at: https://financialtribune.com/articles/people-travel/50038/iranian-ecolodges-increase-tenfold (accessed 27 April 2020).

Financial Tribune (2017) Iran reviewing rural tourism potentials. Available at: https://financialtribune.com/articles/people-travel/70556/iran-reviewing-rural-tourism-potentials (accessed 10 June 2020).

Financial Tribune (2020) Corona closes 2,200 eco-lodges. Available at: https://financialtribune.com/articles/travel/103144/corona-closes-2200-eco-lodges (accessed 2 May 2022).

Getz, D. and Carlsen, J. (2005) Family business in tourism: state of the art. *Annals of Tourism Research* 32(1), 237–258.

Ghaderi, Z. and Henderson, J.C. (2012) Sustainable rural tourism in Iran: a perspective from Hawraman village. *Tourism Management Perspectives* 2–3, 47–54. DOI: 10.1016/j.tmp.2012.03.001.

Giampiccoli, A. and Kalis, J.H. (2012a) Community-based tourism and local culture: the case of the amaMpondo. *Revista de Turismo y Patrimonio Cultural* 10(1), 173–188. DOI: 10.25145/j.pasos.2012.10.017.

Giampiccoli, A. and Kalis, J.H. (2012b) Tourism, food, and culture: community-based tourism, local food, and community development in Mpondoland. *Culture, Agriculture, Food and Environment* 34(2), 101–123. DOI: 10.1111/j.2153-9561.2012.01071.x.

Goodarzi, Z., Abbasi, E. and Farhadian, H. (2018) Achieving consensus deal with methodological issues in the Delphi technique. *International Journal of Agricultural Management and Development* 8(2), 219–230.

Goodwin, H. and Santilli, R. (2009) Community-based tourism: a success? *ICRT Occasional Paper* 11(1), 1–37.

Graci, S.R. (2012) Putting community based tourism into practice: the case of the Cree village ecolodge in Moose factory, Ontario. *Téoros* 31(3), 65–70. DOI: 10.7202/1036565ar.

Han, H., Eom, T., Al-Ansi, A., Ryu, H.B. and Kim, W. (2019) Community-based tourism as a sustainable direction in destination development: an empirical examination of visitor behaviors. *Sustainability* 11(10), 2864. DOI: 10.3390/su11102864.

Haqbin, A., Shojaei, P. and Radmanesh, S. (2022) Prioritising COVID-19 recovery solutions for tourism small and medium-sized enterprises: a rough best–worst method approach. *Journal of Decision Systems* 31(1–2), 102–115. DOI: 10.1080/12460125.2021.1927487.

Hesam, M. and Baghiani, H. (2018) Assessing the attitude of villagers towards the implementation of community-based rural tourism (case study: Gurab Pas, fuman County, Guilan Province). *Journal of Rural Research* 9(1), 72–87.

IANA (2017) Development of sustainable tourism and economic empowerment of local communities through the establishment of eco-lodges. Available at: https://www.iana.ir (accessed 21 August 2020).

IRNA (2016) A network of community lodges has been set up in villages targeted for tourism in Esfahan. Available at: https://www.irna.ir/news/82172180 (accessed 19 February 2021).

Johnson, P. (2010) Realizing rural community-based tourism development: prospects for social economy enterprises. *Journal of Rural and Community Development* 5(1), 150–162.

Juma, L.O. and Khademi-Vidra, A. (2019) Community-based tourism and sustainable development of rural regions in Kenya; perceptions of the citizenry. *Sustainability* 11(17), 4733. DOI: 10.3390/su11174733.

Kalantari, K., Fami, H.S., Asadi, A., Qasemi, I. and Chubchian, S. (2008) Major challenges of Iranian rural communities for achieving sustainable development. *American Journal of Agricultural and Biological Sciences* 3(4), 724–728. DOI: 10.3844/ajabssp.2008.724.728.

Kayat, K. (2014) Community-based rural tourism: a proposed sustainability framework. *SHS Web of Conferences* 12, 01010. DOI: 10.1051/shsconf/20141201010.

Keyim, P. (2017) Tourism collaborative governance and rural community development in Finland: the case of Vuonislahti. *Journal of Travel Research* 57(4), 483–494. DOI: 10.1177/0047287517701858.

Khalid, S., Ahmad, M.S., Ramayah, T., Hwang, J. and Kim, I. (2019) Community empowerment and sustainable tourism development: the mediating role of community support for tourism. *Sustainability* 11(22), 6248. DOI: 10.3390/su11226248.

Korez-Vide, R. (2013) Promoting sustainability of tourism by creative tourism development: how far is Slovenia? *Innovative Issues and Approaches in Social Sciences* 6(1), 77–102. DOI: 10.12959/issn.1855-0541.IIASS-2013-no1-art05.

Kovjanic, G. (2014) Islamic tourism as a factor of the Middle East regional development. *TURIZAM* 18(1), 33–43. DOI: 10.5937/Turizam1401033K.

Krstj, B., Jovanovj, S. and Jankovj-Milj, V. (2008) Sustainability performance management system of tourism enterprises. *Facta Universitatis-series: Economics and Organization* 5(2), 123–131.

Lane, B. (1994) Sustainable rural tourism strategies: a tool for development and conservation. *Journal of Sustainable Tourism* 2(1–2), 102–111. DOI: 10.1080/09669589409510687.

Mahdavi, D., Eftekhari, R.A. and Sojasi Gheidari, H. (2016) Design strategies for the sustainable development of tourism Iranian historic-cultural villages. *Journal of Geography and Planning* 20(56), 275–300.

Mamirkulova, G., Mi, J., Abbas, J., Mahmood, S., Mubeen, R. *et al.* (2020) New silk road infrastructure opportunities in developing tourism environment for residents better quality of life. *Global Ecology and Conservation* 24, e01194. DOI: 10.1016/j.gecco.2020.e01194.

Masaeli, A., Rezaei, N., Derafshi, K. and Basakha, M. (2021) Effects of corona epidemic on labor employment in Iran's tourism-related industries. *Tourism Management Studies* 16(56), 279–308.

MCTH (2020) List of community lodges by province. Unpublished statistics. Iranian Ministry of Cultural Heritage, Tourism and Handicrafts, Tehran, Iran.

MCTH (2021) Community-based tourism criteria in Iran. Available at: https://kurdistan.mcth.ir/boomgardi# (accessed 19 February 2021).

Mearns, K. (2011) Using sustainable tourism indicators to measure the sustainability of a community-based ecotourism venture: Malealea Lodge & Pony Trek Centre, Lesotho. *Tourism Review International* 15(1), 135–147. DOI: 10.3727/154427211X13139345020499.

Mearns, K. (2015) Applying sustainable tourism indicators to community-based ecotourism ventures in southern Africa. *Athens Journal of Tourism* 2(3), 179–194. DOI: 10.30958/ajt.2-3-4.

Mirani, S.Z. and Farahani, B.M. (2015) Second homes tourism and sustainable rural development in all around the world. *International Journal of Leisure and Tourism Marketing* 4(3/4), 176. DOI: 10.1504/IJLTM.2015.072076.

Mirzaei, R., Sadin, M. and Pedram, M. (2021) Tourism and COVID-19: changes in travel patterns and tourists' behavior in Iran. *Journal of Tourism Futures* 1–13. DOI: 10.1108/JTF-01-2021-0017.

Morakabati, Y. (2013) Tourism in the Middle East: conflicts, crises and economic diversification, some critical issues. *International Journal of Tourism Research* 15(4), 375–387. DOI: 10.1002/jtr.1882.

Mtapuri, O. and Giampiccoli, A. (2013) Interrogating the role of the state and nonstate actors in community-based tourism ventures: toward a model for spreading the benefits to the wider community. *South African Geographical Journal* 95(1), 1–15. DOI: 10.1080/03736245.2013.805078.

Mu, E. and Pereyra-Rojas, M. (2017) Understanding the analytic hierarchy process. In: Mu, E. and Pereyra-Rojas, M. (eds) *Practical Decision Making: An Introduction to the Analytic Hierarchy Process (AHP) Using Super Decisions V2*. Springer, Cham, Switzerland, pp. 7–22.

Nematpour, M., Khodadadi, M., Rezaei, N. and Makian, S. (2020) Structural analysis of the development of the Iranian tourism market employing a MICMAC approach: a new long-range planning method to attract the ASEAN international tourist market. *Journal of Hospitality and Tourism Insights* 4(4), 393–417. DOI: 10.1108/JHTI-04-2020-0053.

Nooripoor, M., Khosrowjerdi, M., Rastegari, H., Sharifi, Z. and Bijani, M. (2020) The role of tourism in rural development: evidence from Iran. *GeoJournal* 86(4), 1705–1719. DOI: 10.1007/s10708-020-10153-z.

Novelli, M. and Gebhardt, K. (2007) Community based tourism in Namibia: 'reality show' or 'window dressing'? *Current Issues in Tourism* 10(5), 443–479. DOI: 10.2167/cit332.0.

O'Gorman, K., Baum, T., McLellan, L. and Butler, R. (2007) Tourism in Iran: central control and indigeneity. In: Butler, R. and Hinch, T. (eds) *Tourism and Indigenous Peoples: Issues and Implications*. Routledge, Abingdon, UK, pp. 269–282.

Pakshir, L. and Nasir, V. (2011) Sustainability of homestay as a form of community-based tourism (CBT): a case study of the rural community in Bavanat-Iran. *Journal of Hospitality and Tourism Management* 8(1), 5–18.

Polnyotee, M. and Thadaniti, S. (2015) Community-based tourism: a strategy for sustainable tourism development of Patong beach, Phuket island, Thailand. *Asian Social Science* 11(27), 90. DOI: 10.5539/ass.v11n27p90.

Purbasari, N., Manaf, A., Hadiyanto, Sudarno and Maryono (2018) Comparative study on the characteristics of community-based tourism between Pentingsari and Nglanggeran tourism village, special region Yogyakarta. *E3S Web of Conferences* 31, 09007. DOI: 10.1051/e3sconf/20183109007.

Radioeqtesad (2018) The need to develop community-based lodges. Available at: http://radioeqtesad.ir/NewsDetails/?m=094146&n=245091 (accessed 23 August 2020).

Rajković, I. and Bojović, M. (2016) Revitalization of traditional architecture towards sustainable development of the Skadar lake area. *Architecture and Urban Planning* 11(1), 32–42. DOI: 10.1515/aup-2016-0005.

Rezvani, M. and Azizi, S. (2019) The impact of community lodges on rural development. In: *The Second National Conference on Iran Sustainable Rural Development*, Kharazmi University, Tehran, Iran, pp. 1–9.

Roshan, G., Yousefi, R. and Błażejczyk, K. (2018) Assessment of the climatic potential for tourism in Iran through biometeorology clustering. *International Journal of Biometeorology* 62(4), 525–542. DOI: 10.1007/s00484-017-1462-6.

Sasu, K.A. and Epuran, G. (2016) An overview of the new trends in rural tourism. *Bulletin of the Transilvania University of Braşov–Economic Sciences* 9(2), 119–126.

Sebele, L.S. (2010) Community-based tourism ventures, benefits and challenges: Khama Rhino sanctuary Trust, Central District, Botswana. *Tourism Management* 31(1), 136–146 DOI: 10.1016/j.tourman.2009.01.005.

Seetanah, B., Juwaheer, T.D., Lamport, M.J., Rojid, S., Sannassee, R.V. *et al.* (2011) Does infrastructure matter in tourism development? *University of Mauritius Research Journal* 17(1), 89–108. DOI: 10.4314/umrj.v17i1.70731.

Shamsoddini, A. (2015) Evaluation of tourism status in sustainable rural development by Reynard model (case study: Pa Ghaleh village, Mamasani county, Iran). *Open Journal of Ecology* 05(3), 80–85. DOI: 10.4236/oje.2015.53008.

Sharpley, R. (2007) Flagship attractions and sustainable rural tourism development: the case of the Alnwick garden, England. *Journal of Sustainable Tourism* 15(2), 125–143. DOI: 10.2167/jost604.0.

Sigala, M. (2020) Tourism and COVID-19: impacts and implications for advancing and resetting industry and research. *Journal of Business Research* 117, 312–321. DOI: 10.1016/j.jbusres.2020.06.015.

Stockdale, A. (2004) Rural out-migration: community consequences and individual migrant experiences. *Sociologia Ruralis* 44(2), 167–194. DOI: 10.1111/j.1467-9523.2004.00269.x.

Straka, J. and Tuzová, M. (2016) Factors affecting development of rural areas in the Czech republic: a literature review. *Procedia – Social and Behavioral Sciences* 220, 496–505. DOI: 10.1016/j.sbspro.2016.05.525.

Sutresna, I.B., Suyana, U.I.M., Saskara, I.A.N. and Wiwin, S.N.P. (2019) Community based tourism as sustainable tourism support. *Russian Journal of Agricultural and Socio-Economic Sciences* 94(10), 70–78. DOI: 10.18551/rjoas.2019-10.09.

Taheri, G. and Emamgholi, A. (2017) Tourism improvement based on sustainable architecture and new functions of textures and historical buildings and their alterations. *Bulletin de la Société Royale des Sciences de Liège* 86, 831–844. DOI: 10.25518/0037-9565.7020.

Taleai, M. and Mansourian, A. (2008) Using Delphi-AHP method to survey major factors causing urban plan implementation failure. *Journal of Applied Sciences* 8(15), 2746–2751. DOI: 10.3923/jas.2008.2746.2751.

Taleshi, M. and Khodapanah, K. (2020) Assessing and measuring tourism sustainability in the villages of tourism destination from the viewpoint of local managers (case study: Ardabil district). *Journal of Studies of Human Settlements Planning* 15(1), 267–280.

Tehran Times (2017) Booming small businesses can buttress tourism in Iran. Available at: https://www.tehrantimes.com/news/417646/Booming-small-businesses-can-buttress-tourism-in-Iran-official (accessed 18 July 2020).

Tehran Times (2020) Some 2,000 eco-lodges operating across Iran. Available at: https://www.tehrantimes.com/news/444003/Some-2-000-eco-lodges-operating-across-Iran (accessed 15 July 2020).

Thetsane, R.M. (2019) Local community participation in tourism development: the case of Katse villages in Lesotho. *Athens Journal of Tourism* 6(2), 123–140. DOI: 10.30958/ajt.6-2-4.

Wang, X., Ren, H., Wang, P., Yang, R., Luo, L. *et al.* (2018) A preliminary study on target 11.4 for UN Sustainable Development Goals. *International Journal of Geoheritage and Parks* 6(2), 18–24. DOI: 10.17149/ijgp.j.issn.2577.4441.2018.02.002.

Yamashita, S. (2011) Community-based associations for sustainable tourism development: fostering sustainable development in developing countries. MSc thesis, Uppsala University, Sweden.

Yanes, A., Zielinski, S., Diaz Cano, M. and Kim, S. (2019) Community-based tourism in developing countries: a framework for policy evaluation. *Sustainability* 11(9), 2506. DOI: 10.3390/su11092506.

Yarihesar, A. and Bakhtar, S. (2016) Application of Alen Prescott stability measurement model in analysis rural sustainable tourism. case study: Nir county-Ardabil Province. *Regional Planning* 6(22), 121–134.

Yousefi, K. (2016) The impact of rural tourist recreational space design on tourism development and tourist attraction in Iran. *Online Journal of Design, Art and Communication* 6, 1432–1437. DOI: 10.7456/1060AGSE/027.

Yu, C.P., Cole, S.T. and Chancellor, C. (2016) Assessing community quality of life in the context of tourism development. *Applied Research in Quality of Life* 11(1), 147–162. DOI: 10.1007/s11482-014-9359-6.

Zhang, H., Sekhari, A., Ouzrout, Y. and Bouras, A. (2014) Deriving consistent pairwise comparison matrices in decision making methodologies based on linear programming method. *Journal of Intelligent & Fuzzy Systems* 27(4), 1977–1989. DOI: 10.3233/IFS-141164.

Zielinski, S., Jeong, Y., Kim, S. and B. Milanés, C. (2020) Why community-based tourism and rural tourism in developing and developed nations are treated differently? A review. *Sustainability* 12(15), 5938. DOI: 10.3390/su12155938.

5 Heritage Conservation for Tourism Development: Identifying the Challenges in Developing Countries – The Case of Iran

Fatemeh Bagheri[1,2]*, Ali Hajinejad[1] and Naser Abdi[2]
[1]University of Guilan, Rash, Iran; [2]Research Center for Tourism, Sustainability and Well-being (CinTurs), University of Algarve, Faro, Portugal

5.1 Introduction

The importance of heritage for a nation's identity (Forster and Kayan, 2009) and cultural/historical tourism development (Kisusi and Masele, 2018) highlights the necessity for proper conservation and management. The concept of heritage management has arisen in response to the special needs and challenges of heritage. Heritage conservation is a reference point to look at the past, understand the present and plan for the future (Snyder, 2008). Also, heritage assets represent the special resources for tourism growth, thus 'conservation is a vital component of their management' (Millar, 1989, p. 9).

Indeed, cultural heritage tourism destinations need to learn how to maintain, develop and utilize their distinctiveness, authenticity and local sense of identity. It is the only way that tourists' activities can be culturally (and economically) sustainable (Domšić, 2013).

Conservation of heritage tourism assets needs multiple factors and actors working together to achieve a common goal. There are various barriers to conservation in most countries. Particularly in developing countries, the governments' pursuit of modernization, population growth, rapid growth in the size of cities (Steinberg, 1996; Throsby, 2016), and problematic political and socio-economic structures are some of the main threats to heritage conservation in developing countries.

Nevertheless, only a few studies have tried to address heritage conservation challenges in developing countries, analyse extant problems and identify the main causes as the first steps in the problem-solving process (Franco *et al.*, 1997). In other words, as long as there is no clear identification of heritage conservation challenges, no effort to deal with such challenges will be effective and any resources allocated will, most likely, be wasted.

Among 193 countries, with 24 subscripted heritage sites, Iran was ranked 10th in the United Nations Educational, Scientific and Cultural Organization (UNESCO) ranking in 2019 (CHTM, 2020). According to the Iranian Cultural Heritage, Tourism and Handicraft Ministry (hereafter, CTHM), which is the legal entity responsible for heritage conservation and policy making in Iran, more than 30,000 historic sites have been completely recorded in the national heritage list (CHTN, 2020). There is a long history of visits to these monuments

*Corresponding author: f.bagheri@guilan.ac.ir

DOI: 10.1079/9781789249149.0005

by both international and domestic tourists in different cities of Iran (Roque and Forte, 2019).

In other words, cultural heritage tourism is one of the oldest types of tourism development in Iran (Mikaeili and Aytuğ, 2019). Although the UNESCO - inscribed heritage of Iran receives sufficient financial and managerial attention from international and national organizations, in many smaller Iranian cities, there are remarkable samples of architecture associated with the past (relevant to both ancient and Islamic heritage) which seem to be neglected by the tourism industry. While heritage conservation could be considered as an opportunity for tourism growth in remote towns and cities, the conservation condition of many Iranian monuments is so poor that they are endangered (Misra, 2009; Rouhani, 2011).

As a result, the need to work on heritage conservation challenges for developing countries in general, and Iran as a developing country in particular, emerges as significant. This study examines practitioners' and scholars' opinions in an attempt to develop a theoretical framework which can portray the heritage conservation challenges in Iran. The chapter starts with a literature review that reveals the need to work on heritage conservation challenges. The chapter then presents the methods undertaken to address this gap in the case of Iran, and describes the data collection process. Finally, the chapter discusses the findings of this study and concludes with specific scholarly implications and implications relevant to both Iranian stakeholders and those in other developing countries.

5.2 Literature Review

5.2.1 Heritage and tourism

Heritage has become an important field of academic study and research (De La Torre, 2013). 'Heritage sites provide tangible links between the past, the present, and the future' (Millar, 1989, p. 9; Günlü *et al.*, 2009), illustrate a strong visual document of regional history and also comprise the basis of tangible local culture (Aziz, 2005). Therefore, the value of heritage sites increases over time. Natural and environmental factors (e.g. humidity, rainwater, flood, earthquake, etc.) and human factors (through development and over-tourism or deliberately destructive actions) further support the need to adequately conserve heritage and historic sites.

Indeed, conservation refers to any process of 'looking after' a place in order to retain its historical, architectural, aesthetic and cultural significance (Hall, 2013) or 'the process of understanding, interpreting and managing the architectural heritage in order to safely deliver it to posterity' (Hegazy, 2015, p. 264). It leads to the durability of a country's heritage and identity symbols (i.e. socio-cultural benefits) (Forster and Kayan, 2009; Kisusi and Masele, 2018). Also, conservation provides a basis to attract historic tourists, boost growth in the tourism industry, and create jobs and vocational-training opportunities (i.e. economic benefits) (Johar *et al.*, 2011; Kisusi and Masele, 2018).

Cultural heritage tourism is one of the social and economic functions of cultural heritage (Doganer and Dupont, 2013) and has recently been re-affirmed by the UNWTO as a major element of international tourism consumption across the world, accounting for over 39% of tourism arrivals (Richards, 2018). Cultural heritage tourism development is also a great strategy for raising the awareness of local communities to create sustainable economic growth (Matiz *et al.*, 2012). In fact, 'when heritage tourism is done right, the biggest beneficiaries are not visitors but local residents who experience a renewed appreciation and pride in their local city as well as its history' (Kisusi and Masele, 2018, p. 117).

However, the relationship between tourism development and heritage conservation has been a controversial issue (Van Zyl, 2005). Towards the end of the 19th century, the first signs of a rapidly developing tourism industry suggested a possible threat to natural and cultural heritage sites (Dabrowski, 2007). This threat is particularly evident when tourist flows overtake the site's carrying capacity (Matiz *et al.*, 2012). As a result, several scholars have investigated the negative impacts of tourism development on heritage sites (e.g. Patuelli *et al.*, 2013; Al-Saad, 2016), compelling even some destinations in developed countries to minimize or control tourism growth (such as the Golden Triangle area in Bruges Historic Centre, Belgium, and the Cinque Terre area in Liguria, Italy, to name a few (Panteia, 2015).

Generally, increased visitor pressure and congestion, overloaded infrastructure and the detriment of the landscape, irresponsible behaviour from tourists and air/ noise/visual pollution are the major negative impacts of tourism development on heritage sites (Farid, 2016; Boudiaf, 2019). However, 'culture and tourism have always been inextricably linked' (Greg, 2018, p. 2) and many researchers (e.g. Henderson, 2002; Van Zyl, 2005; Uddhammar, 2006) believe that adequate, well-planned management can act as a catalyst to create a mutual, advantageous link between tourism development and heritage conservation.

5.2.2 Challenges to heritage conservation in developing countries

Heritage conservation in developing countries faces additional challenges. Aziz *et al.* (2016) identified 46 main challenges to heritage conservation in Malaysia and classified them into five themes, namely technical, environmental, organizational, financial and human issues.

More recently, Mubaideen and Al Kurdi (2017) presented a supporting management model for the effective incorporation of archaeological sites in the planning process in Jordan. Here, the main challenges of cultural heritage conservation are organized into five categories: (i) ineffective conservation legislation; (ii) lack of clearly defined prioritization mechanisms to absorb the diversity of distinctive values and qualities of cultural heritage sites; (iii) inadequate reinforcement of the present law because of the lack of awareness of private and public institutions; (iv) absence of technical specialists; and (v) lack of efficient cohesion and insufficient cooperation and coordination between different related institutions.

Concerning the last challenge mentioned, Amar *et al.* (2016, p. 227) and Jamieson (2000) explicitly stated that 'to develop sustainable tourism in heritage areas, all stakeholders, especially those at the community level, must learn to work together cooperatively'. Amar (2017) believed that the clear allocation of roles and well-defined responsibilities among stakeholders are required to promote the achievement of conservation goals and sustainability outcomes of the heritage conservation.

Also, Aas *et al.* (2005) argued that to minimize threats caused by the conflicts between tourism and heritage groups, there is a need for dialogue, cooperation and collaboration among the various stakeholders involved. This implies the need for communication among the tourism sectors (tourism management departments, tourism investment companies, tourists, travel agencies), local residents and committees, expert groups and administrations of architectural heritage protection, construction companies, media and government (central and local) (Wang *et al.*, 2019).

In particular, many studies (e.g. Yung and Chan, 2011; Macdonald and Cheong, 2014; Choenni, 2015; Spiridon and Sandu, 2016; Jokilehto, 2017; Wong, 2018) emphasized that the local community should be involved in heritage conservation to enhance the quality of planning, to reduce the likelihood of conflict and to increase the community's ownership of its heritage (Hall and McArthur, 1998). In contrast, tokenism is only a superficial exercise whereby community participants become merely information providers; at best they get involved in consultation and not in more important and effective positions with decision-making power (Arnstein, 1969). However, 'in many developed countries, participatory approaches to heritage conservation and management have proven to be successful and more culturally and socially sustainable than top-down approaches' (Wong, 2018, p. 18).

As mentioned before, the engagement of professional, trained people in the conservation process is clearly vital (Ahmad, 2013; Al-Tokhais and Thapa, 2019). The lack of conservation specialists and personnel training emerges as the biggest challenge of conservation projects in Malaysia (Aziz *et al.*, 2016). Also, Bello *et al.* (2012) concluded that lack of human and financial resources was very prevalent in heritage rehabilitation plans across various municipal projects in Brazil and Malaysia. Similarly, Kammeier (2008) clearly asserted that in poor countries, professional potentials are as limited as financial resources. Confirming these results, Harun (2011, p. 52) highlighted 'the lack of skilled workers and responsibility of conservators' as the major problem in the heritage conservation of Malaysia.

In addition to the problems that may arise due to the relationships between respective organizations and community participation, a more fundamental issue relates to the laws and legal considerations concerning heritage conservation. Research on heritage conservation emphasizes that standard protection is a decisive factor in heritage conservation (for details, see Fitri *et al.* (2016) on Malaysia, Otieno (2015) on Kenya, Vásquez (2009) on Mexico and Eboreime (2008) on African countries).

In other words, in this context, it seems that the first role of governments is enacting laws that promote and encourage the conservation of historic sites and monuments that are deemed to help the current and future generations to understand the histories of the local area, state or territory as well as federal environments (Amar, 2017). Overall, transparent, updated legal frameworks and enforcement of relevant national/international laws can be among the protective and preventive measures for heritage conservation. Nonetheless, laws cannot be effective without any imperative measures – a process that ensures compliance with legal requirements (Otieno, 2015).

Some studies refer to other challenges in heritage conservation for cultural tourism development related to the protective affairs of heritage. Generally, most historic sites and buildings are located in the old town of cities where urban development plans and policies (including transportation and other infrastructure) prioritize renovation and modernization. In other words, cities and towns in developing countries increasingly face the challenge of balancing urban conservation and development due to rapid urbanization and development pressure (Leitao, 2011; Puren and Jordaan, 2014). 'The conflict between conservation and modernization is deeply rooted in the history of urban development' (Martínez, 2017) and is not a recent issue.

5.2.3 Heritage tourism development in Iran

Located in the Middle East, Iran has an ancient civilization and numerous historic and natural historic sites in different towns and cities. Figure 5.1 illustrates all Iranian sites that have been designated as World Heritage Sites and reveals that 22 out of 24 of Iran's World Heritage Sites are archaeological ones.

Still, thousands of lesser-known historic places lie in different areas and cities of the country and date back to the Islamic and pre-Islamic (especially the Achaemenes and Sassanid Empires) eras. To safeguard these cultural assets, Iranian heritage experts have provided the National Heritage List as an effective legal instrument (Rouhani, 2011). Despite these measures and efforts for the heritage conservation and restoration of Iran's history, 'the development of heritage has been a subject of debate as a great number of ancient sites in Iran belong to the pre-Islamic era' (Seyfi *et al.*, 2018, p. 4). These conflicts, along with other challenges that this chapter discusses, push heritage conservation into a critical condition.

Its numerous attractions prompt millions of international and domestic tourists to visit Iran annually. According to UNWTO highlights, just before the pandemic, in 2019 alone, 9,107,000 international tourists visited Iran (UNWTO, 2020). Among the various types of tourism growth, in the pre-COVID-19 era, cultural heritage tourism was one of the most prevalent types of tourism in Iran with a large number of international tourists travelling in order to visit ancient and Islamic archaeological sites. Despite the popularity of the classic route in Iran (Tehran–Esfahan–Shiraz) among tourists, a large number of monuments and historic sites are located in other Iranian cities.

Contradictions between heritage conservation and tourism growth often lead to conflicts, and have simultaneously attracted significant concern from the general public in Iran (Seyfi *et al.*, 2018). Generally, the maintenance of Iran's World Heritage Sites is according to the worldwide guidelines. However, despite the existence of heritage tourism development potential in many cities of Iran (especially in Qazvin, Sistan and Baluchistan, Hamedan, Kermanshah, Eastern and Western Azerbaijan, etc.), conservation efforts in these cities appear to be limited. For instance, Qazvin city – which was the capital of Iran during the Safavieh Empire (1524–1580) – hosts a large number of archaeological sites.

Yet municipal projects led to the destruction of several historic buildings throughout the city (such as the historic houses in the Bolaghi district) (Imanian, 2019). In another case, three

Fig. 5.1. World Heritage Sites of Iran. Figure used with permission from Persia Advisor Travel (2017).

historic houses in the Kababian district, located in the city of Hamedan, were demolished by the municipality due to urbanization projects (*Isna News*, 2017). Such examples suggest that tourism priorities present only part of the problem in Iran.

5.3 Research Methodology

5.3.1 Collection of textual information

This study uses an exploratory approach to identify and categorize the challenges of heritage conservation for tourism development in Iran. Towards this end, analysis of online information took place and the sample of websites was selected through a comprehensive search of website lists under the directories of Yahoo and Google from 1 November 2019 to 1 March 2020. In this step, the researchers used three main keywords to search on these search engines: 'problems of the heritage/historic sites in Iran', 'challenges of the heritage/historic sites in Iran', 'condition of heritage/historic sites in Iran'. Using these keywords, more than 100 websites (in Persian) were found. The subsequent analysis included only websites with specific relevance to the research question and excluded more generic pages.

In the next step, 45 related websites were classified into three subcategories: 'tourism-related websites', 'news websites (magazines)' and 'governmental/official websites'. Most of these websites contained the text of Iranian officials' and authorities' interviews about heritage conservation. Table 5.1 presents a comprehensive list of the websites included in our analysis. Among the reviewed websites, there were 3 tourism-related websites, 35 news websites (magazines) and 6 governmental official websites.

To analyse the data, the content of the websites was first saved into separate Word files (.doc files) and the text data were then merged into three separate files for every subcategory for further content analysis. Then, the textual information of 45 websites was studied carefully and NVivo 8 software was used for quantitative thematic analysis. Each file was read several times to allow the researchers' immersion in the data set.

5.3.2 Coding and analysis

Thematic analysis is a foundational method for qualitative analysis (Braun and Clarke, 2006) and a method to identify, analyse and report patterns within a data set. Note that there is some controversy on what it is and how it should be performed (Tuckett, 2005). In any case, once the text was entered into the software and studied several times, three steps were followed.

Development of basic themes: In the first step of thematic analysis, the coding of all collected data was done line by line, so as not to miss any important information. It is essential to ensure the reliability and validity of the coding by researchers, so some textual information was coded by another researcher familiar with the coding process and any disagreements were resolved.

Consolidation into organizing themes: In the second level of coding, the main aim was classifying and summarizing the basic themes (the result of the previous section) into fewer categories. Indeed, in this phase, the basic themes were clustered into coherent groups and condensed into higher-level themes that shared similarities (Walters, 2016).

Deriving overarching global themes and networks: In the final step, a deeper analysis of the organizing themes at a higher level of abstraction revealed overarching global themes which could be derived from them, while maintaining their internal homogeneity and mutual exclusivity (Braun and Clarke, 2006). 'The global theme could be seen as the conclusion or main tenet of an argument' (Walters, 2016, p. 114). After extracting all themes in three levels, a web-like diagrammatic network was drawn providing a valuable graphic representation of the themes (Attride-Stirling, 2001).

5.4 Findings

Table 5.2 presents the results of the thematic analysis of the websites related to the heritage conservation challenges for Iran's cultural heritage tourism in three levels (basic, organizing and global). Figure 5.2 presents a graphic representation of the thematic network consisting of three levels of data analysis (18 basic themes, 4 organizing themes and 2 global themes) as one of the outputs of NVivo 8 software. Basic, organizing and global themes will be examined in the following section.

5.4.1 Basic themes

Through a precise analysis of the collected data, 376 references and 18 basic themes were extracted in the first level. Table 5.2 indicates that the highest frequency (n = 38; 10.1%) was related to the basic theme of 'budget shortfall' and financial limitation is one of the major issues in the heritage conservation of Iran. A governmental representative has actually asserted: 'In this context, we have wide-range duties and if we only want to exercise our obligations on the national and world recorded monuments, we will need at least 16 trillion Tomans.'

'Distrust/lack of coordination between the organizations' with 32 occurrences (8.5%) was identified as the second major challenge. Because of the variety of the responsible heritage organizations in Iran, many believe that there is

Table 5.1. List of websites.

Websites	Web address	Geographical focus
Tourism-related websites	aitotravel.ir	General
	chamedanmag.com	Esfahan city
	sahelsefid.com	General
	donyayesafar.com	Khorasene-e-razavi province
Governmental/official websites	namadeghtesad.ir	General
	ashtian.ostan-mr.ir	Ashtian city
	mizanonline.com	Yazd city
	amordadnews.com	General
	sedayemiras.ir	General
	hamshahrionline.ir	General
	donya-e-eqtesad.com	General
	mehrnews.com	General
	mashreghnews.ir	General
	mokhbernews.ir	Chaharmahal and Bakhtiari province
	qudsonline.ir	General
	shafaqna.com	General
	shoaresal.ir	General
	sinapress.ir	General
	saat24.news	General
	rouydad24.ir	General
	dana.ir	Mazandaran province
	fararu.com	General
	qazvinbar.org	Qazvin province
	jamaran.ir	General
	khabarazma.com	Hamedan province
	khorasannews.com	Sistan and Baluchistan province
	entekhab.ir	Sistan and Baluchistan province
	parsnews.com	Tehran city
	yjc.ir	General
	imannews.com	General
	newspaper.hamshahri.org	General
	imna.ir	Esfahan city
	ilna.ir	General
	iran-newspaper.com	Semnan city
	ion.ir	General
	tasnimnews.com	Kordestan province
	tabnak.ir	General
	isna.ir	Khaf city

Continued

Table 5.1. Continued

Websites	Web address	Geographical focus
	titreshahr.com	General
News websites	iribnews.ir	General
	ichto.ir	General
	news.shiraz.ir	Shiraz city
	fa.irunesco.org	General
	news.mrud.ir	General
	icana.ir	General

Table 5.2. The coding spectrum of collected data.

Global themes	Organizing themes	Basic themes	Frequency (n)	Percentage (%)	Rank
Soft challenges	Organizational issues	Structural problems	20	5.3	9
		Unclear responsibility	26	6.9	4
		Bureaucracy in organizations	11	2.9	13
		Distrust/lack of coordination between the organizations	32	8.5	2
		Negative attitude to heritage and (urban/rural) development	21	5.6	8
	Local people issues	Weak/lack of local awareness	27	7.2	3
		Weak/lack of introduction of heritage to locals	22	5.9	7
		Weak/lack of local participation	24	6.4	5
Hard challenges	Technical issues	Construction in heritage surroundings	23	6.1	6
		Improper exploration methods	15	4.0	12
		Improper conservation mechanisms	17	4.5	11
		Budget shortfall	38	10.1	1
		Lack of security force presence	18	4.8	10
		Lack of professional human resources	26	6.9	4
	Legal issues	Weakness in the national recording process	23	6.1	6
		Weak conservation laws	18	4.8	10
		Not enforcing the laws	10	2.6	14
		Weakness in transferring laws of heritage to private sector	5	1.3	15

not enough cooperation and trust between these organizations to conserve the historic sites and promote them for tourism development. This was one response:

> The deputy of cultural heritage attributes the problems of the cultural heritage to the conflict between the involved organizations... The cultural heritage researchers also believe

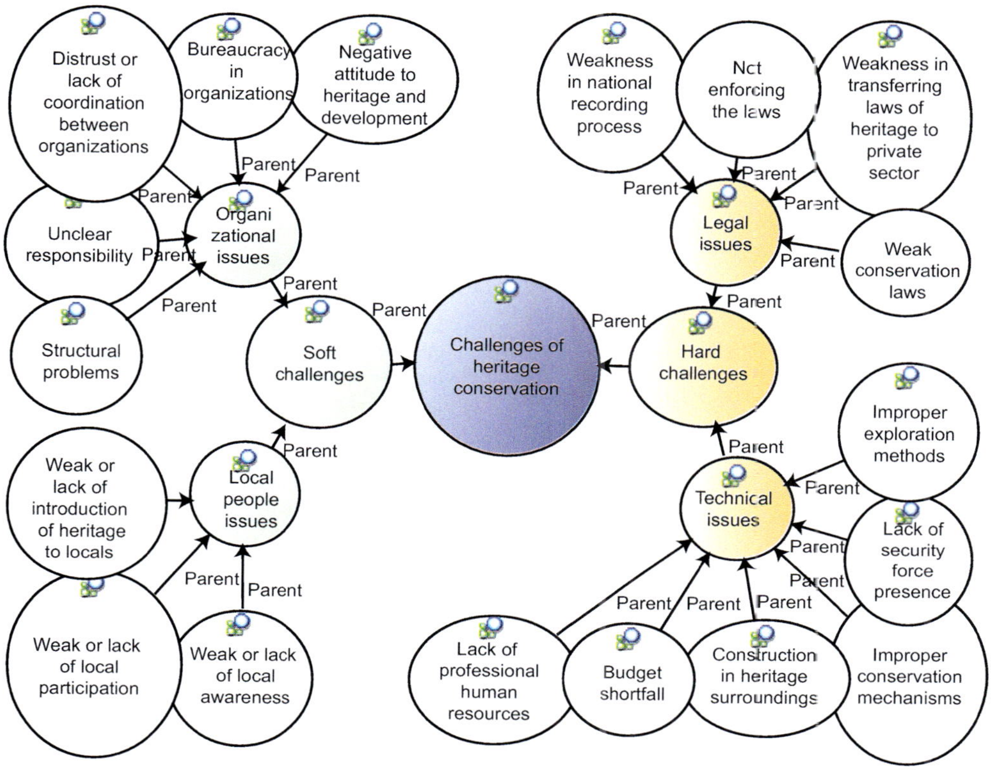

Fig. 5.2. Thematic network of basic, organizing and global themes.

> that there should be a situation that different organizations, such as the municipality, city council, parliament and the CTHO coordinate with each other in the cultural heritage conservation and exploitation.

The third challenge was 'weak/lack of local awareness' accounting for 7.2% of data (n = 27). Undoubtedly, awareness of local people affects their attempts at conserving the heritage which constitutes a major part of their past and identity. The general manager of the historic buildings and districts preservation and restoration of CTHO has asserted that: 'Lack of people's awareness and knowledge is an important barrier to heritage conservation.'

'Unclear responsibility' and 'Lack of professional human resources' with 26 occurrences (6.9%) claimed the fourth rank. Because of the aforementioned variety of related organizations in the context of heritage, there is no transparency related to their responsibilities. One of Tehran's council members has said that: 'There is an old challenge between the government and urban managers on assigning heritage conservation tasks to the councils and municipalities with all its powers and budgets.'

Dealing with professional human resources as a highly effective issue, the chief of the CTHO of Charmahal and Bakhtiari province has remarked that: 'Our organization does not have sufficient human resources and specialized workforces for heritage conservation.'

The fifth challenge was 'Weak/lack of local participation' (n = 24; 6.4%). The importance of this issue, which can be considered as one of the consequences of the third challenge (Weak/lack of local awareness), has been emphasized by several interviewees. The educational deputy of conservation and restoration faculty of Esfahan Art University asserted that: 'In Iran, there is an expectation that government organizations should be in charge of any heritage while in other parts of the world, this support is provided by people, especially the local people.'

'Construction in heritage surroundings' and 'Weakness in the national recording process' were both considered as the sixth challenge (n = 23; 6.1%). One of the problems is damage to the historic sites that occurs, unfortunately, because of failure in the determination of buffer zones and the fact that these historic sites are located on agricultural lands, which leads farmers to enter these historic sites beyond their land and farms.

Concerning the national recording, the chief of Iranshahr city CTHO said that: 'There are many monuments that have been recorded in the national list but they were refused by the Administrative Justice Court (as a national asset) and eventually were damaged.'

'Weak/lack of introduction of heritage to locals', 'Negative attitude to heritage and development' and 'Structural problems' are respectively in the seventh, eighth and ninth positions (n = 22; 5.9%, n = 21; 5.6% and n = 20; 5.3%). Connected with the importance of introducing the heritage to people, a tourism expert notes: 'There are some derelictions even in introducing valuable heritage ... this information can be presented for introducing historic sites and for the participation of people in their identity maintenance.'

Concerning the relationship between urban development and the historic sites, the deputy of urban and rural development affairs of the country's ministry explains that: 'We are witnessing the destruction of old, cultural sites to meet some urban needs and unfortunately, we could not meet them in the right way.'

The subsequent themes shown in Table 5.2 are: 'Weakness in the conservation laws' and 'Lack of security force presence' (n = 18; 4.9%). Concerning the conservation laws, the country's cultural heritage deputy asserts that: 'There are different interpretations of the conservation laws in different organizations.'

On the importance of security forces' presence as a preventive factor, the commander of the cultural heritage conservation of Sistan and Baluchistan province explains that:

> There is a lack of facilities and a lack of the presence of security forces. Further, since several museums of historic sites are established in the province, caring for all of the mentioned cases is not possible owing to this lack of security forces.

'Improper conservation mechanisms' (n = 17; 4.5%) has been ranked as the 11th challenge. Improper conservation mechanism is considered as a technical challenge in heritage conservation according to some different experts. For example, a cultural heritage expert of the CTHO of Tehran province states that:

> In repairing works in the first district of Tehran city, we do not accept the use of Kazakh bricks, but we emphasize the use of this kind of brick in the Oudlajan district. Unfortunately, unlike the common architecture type in this area, even composites exterior design are used. Buildings should be made in accordance with the dominant conditions.

The next issue is related to 'Improper exploration methods' (n = 15; 4.0%). It has been stated that the exploitation methods of heritage and historic sites, especially where these buildings have private owners, is a serious issue for their conservation.

Sometimes the slow procedure of pursuit of the heritage conservation affairs by the CTHO in courts leads to their gradual destruction. In this regard, 'Bureaucracy in organizations' (n = 11; 2.9%) has been the 12th challenge of heritage conservation in Iran.

Concerning 'Not enforcing the laws' (n = 10; 2.6%), as the next challenge, it has been said that regrettably, despite the existence of the laws and technology, the laws have lost their executive effect and today we witness the destruction or violation of cultural heritage. Definitely, we should not blame the law. The protectors should be careful that no organization should neglect law enforcement since *encroachment* onto cultural heritage sites is a kind of violation of public security.

Finally, concerning the challenge of 'Weakness in transferring laws of heritage to the private sector' (n = 5; 1.3%) existence of clear, precise laws and contracts is of absolute necessity. In this regard, the chief executive of the monument's preservation and revival fund in CTHO notes that: 'Some valuable historic sites have been made available to the private sector without any detailed investigations. Also, some filing cases have been made without the applicant's technical and financial eligibility.'

5.4.2 Organizing themes

To summarize and present a comprehensible classification of the 18 basic themes, as Fig. 5.2 shows, these themes have been categorized into four organizing themes based on rational connections between them, researchers' knowledge and some similar studies' results especially Mubaideen and Al Kurdi (2017), Fitri *et al.* (2016), Aziz *et al.* (2016) and Della Torre (2010).

Accordingly, four organizing themes were extracted wherein five themes of 'structural problems', 'unclear responsibility', 'bureaucracy in organizations', 'distrust/lack of coordination between the organizations' and 'negative attitude to heritage and (urban/rural) development' have been categorized into the organizing theme of 'organizational issues'.

Three themes of 'weak/lack of local awareness', 'weak/lack of introduction of heritage to locals' and 'weak/lack of local participation' were placed under 'local people issues' as the other organizing theme. Also, the organizing theme of 'technical issues' was dedicated to six basic themes: 'construction in heritage surroundings'; 'improper exploration methods'; 'improper conservation mechanisms'; 'lack of security force presence'; 'budget shortfall'; and 'lack of professional human resources'. Finally, 'weakness in the national recording process', 'weak conservation laws', 'not enforcing the laws' and 'weakness in transferring laws of heritage to the private sector' because of their interrelationships were grouped into 'legal issues'.

5.4.3 Global themes

In the first column of Table 5.2 and in the centre of Fig. 5.2, there are two global themes: 'hard challenges' and 'soft challenges'. Indeed, in the highest level of data analysis, four organizing themes (i.e. organizational issues, local people issues, technical issues and legal issues) were innovatively divided into these two overarching themes. Consequently, organizational and local-related issues have been classified as 'soft challenges', as these challenges depend on the behavioural aspects and attitudes of authorities and local people to heritage and conservation. However, technical and legal issues have a harder aspect and need more sophisticated, non-behavioural tools. Nonetheless, the soft and hard issues are interrelated with each other and cannot be considered as completely separate issues.

5.5 Discussion

The proposed conceptual framework of the identified challenges in this chapter is comprehensive and creative in most parts compared to the previous studies. In fact, in the majority of the previous studies (such as Jamieson, 2000; Hegazy, 2015), the authors have generally discussed the conservation heritage challenges without providing an integrated model/framework to address and map the challenges. Some others have tried to highlight a specific issue related to conservation; for instance, Yung and Chan (2011), Dian and Abdullah (2013) and Penna *et al.* (2013) discussed issues in public participation and shared management in heritage conservation, and Appendino (2017), Martínez (2017), and Andalib and Abdolahzadefard (2013) investigated the linkage between conservation and urban development plans.

The authors of this chapter have received help from the literature to name the challenges. In the most similar research, Aziz *et al.* (2016) identified five issues on Malaysia's historical sites based on the literature review. Yet, the researchers did not address and discuss some vital issues such as locals' participation/awareness or legal challenges.

The first principal challenge was the budget shortfall. Financial limitations are the major issue in supporting heritage conservation projects (Worthing and Dann, 2000; Kenny, 2006). Since conservation costs a lot but the CTHM's budget is limited, it is to be expected that this is considered one of the serious obstacles in Iran.

Logically, financial problems can directly affect some of the other challenges and intensify them. For instance, the use of professional workers and security teams will diminish or conservation methods will not be as effective as they could be. Some of the interviewees asserted that since the limited budget of the CTHM must be legally divided into three parts of this organization (i.e. cultural heritage, handicrafts and tourism), the heritage department, the major organization responsible for heritage conservation in Iran, always suffers from lack of sufficient budget.

A frequent solution to solve this challenge in many countries is adopting an economic approach and interpreting heritage as cultural capital (Throsby, 2016). Indeed, while it is true that heritage conservation must usually be initiated and managed by governmental authorities, the finance for investment projects could frequently come from both private and public sources in the form of shared projects and long-term public–private partnerships in different shapes.

Also, an organizational weakness is that managers prioritize urban or rural development over heritage conservation. In many provinces, where most historic buildings are located in the old town, some urban development projects, including the restoration of old town or transportation infrastructures, have damaged the historic districts and buildings.

Regrettably, some managers call these valuable parts of cities 'timeworn districts' and do not believe in preserving the city's identity, its socio-cultural value and its attractiveness for cultural tourists. Nevertheless, Martokusumo (2004) and Martínez (2017) argued that urban development can be achieved under the supervision of heritage conservation and the dual nature of these fields creates opportunities that can expand the cultural experience of people. In this way thoughtful and legal integration of heritage conservation into urban development plans (Udeaja *et al*., 2020) can be a really effective solution.

The findings also demonstrate that poor coordination among the involved institutions in heritage conservation is a serious challenge for the protection of historic sites. In Iran, municipality and city council along with CTHOs are the two main relevant contributors to conservation.

Still, in most cases, unclear separation of responsibilities, administrative bureaucracy and the difference in opinions and attitudes of these organizations' executives compound this poor coordination. This is in contrast to the fact that sustainable heritage conservation requires all related organizations to learn how to deal with conflicts and to explore the creative use of partnerships as a way of sharing knowledge (Jamieson, 2000). Therefore, it is vital to encourage stronger cooperation and collaboration among the various stakeholders of heritage conservation (Hall, 2006; Duval and Smith, 2013; Timothy and Boyd, 2015).

In Iran, it seems that the lack of coordination and trust between the mentioned organizations has neutralized the effectiveness of their efforts for conservation. Defining responsible institutions to ensure more effective and articulate communications and actions, promoting incentives directed principally at institutional strengthening (Penna *et al*., 2013) and defining legal structures to create integrated coordination would help to overcome this challenge.

The other significant weakness of heritage conservation in Iran is poor public awareness about the importance of heritage. Despite the locals' sense of belonging to their past and the importance of residents' awareness in the conservation process (Munjeri, 2004), these matters have remained neglected in many less-developed societies. People should first be informed properly to accept and undertake some roles in heritage conservation. Therefore, public education and informative programmes could be effective to raise public awareness and motivate them to participate. In other words, 'the growing awareness of the communities about their tasks in this regard highlights the recognition that governments are not solely responsible for conservation' (Macdonald and Cheong, 2014, p. 12).

In many developing countries, engagement mechanisms remain theoretical or are practically ineffective to allow communities to have meaningful participation. Undoubtedly, choosing the right community participation mechanisms depends on many variables including the nature of historic sites (Wong, 2018) and the public's power to influence decision making (Dian and Abdullah, 2013). However, as Tosun (2005, p. 339) said: 'enacting legal measurements, restructuring administrative systems at the operational level and the actual community consultation or participation process' are the main phases to engage communities in the tourism development process.

In addition to the technical, organizational and community-based challenges, some legal gaps and conflicts were complicating heritage conservation in Iran. The national recording process of historic sites is one of these. As an example of this issue, in recent years, more than 60 historic houses have been removed from the national assets list since their private owners have complained about the CTHO to the Administrative Justice Court of Iran.

According to the first article of the adopted laws of 2011 in Iran's parliament: buildings and places that were constructed before the end of the Zandieh era (the 18th century) can be considered as national assets and are under government supervision. Referring to this law, private owners whose property's history dates back after the Zandeih period do not allow the government to record their properties as a national asset.

Concerning this challenge, Ndoro (2008) believes that it is crucial to know how the law defines the heritage to be protected and it is rarely comprehensive enough to cover all aspects of heritage in one legal instrument. However, the review and harmonization of laws are necessary to manage heritage effectively. While a basic protective law can provide a consolidated framework for other challenges related to cultural heritage and tourism, the laws must be reviewed to ensure harmonization with the objectives and principles of the cultural policy and tourism development (Otieno, 2015).

The other controversial aspect of legal gaps in Iran was the reuse of historic buildings (especially old houses and caravanserais (roadside inns) by private investors to run tourism businesses. The exploitation of these valuable buildings in the form of traditional accommodation or restaurants can be a double-edged sword. It suggests that these tourism-related facilities are truly attractive elements for cultural tourists at a destination (such as Yazd, Kerman and Fars provinces) and create economic value for communities. However, there is the possibility of damage to these properties as long as they are used by exploiters and tourists (customers).

This issue should be considered in legal frameworks and in the implications of contracts between responsible organizations and investment applicants. Continuous monitoring of the implementation of investors' commitments to responsibly maintaining historic buildings and law enforcement are both equally vital.

5.6 Conclusion

Overall, the purpose of this chapter is to identify and classify the challenges of heritage conservation in Iran as a developing country. Generally, few studies have previously tried to map the challenges of historic sites (such as Della Torre, 2010; Aziz *et al.*, 2016; Fitri *et al.*, 2016; Mubaideen and Al Kurdi, 2017). Based on the findings, relevant challenges can be classified into 18 basic themes, 4 organizing themes and 2 global themes. The soft challenges are related to various aspects of the responsible institutions (including CTHOs and municipalities) and local communities, while the hard challenges consist of different technical and legal issues.

With a more precise examination of this study's general themes (hard and soft challenges), it can be understood that these challenges are interrelated and influenced by each other. The challenges, which are found and discussed in this study on conservation of Iranian heritage, are also seen in many other countries, especially less-developed societies due to their harsh socio-economic and political situations (Penna *et al.*, 2013). Certainly, plans and policies must be formulated and implemented to improve the status of these recognized fundamental challenges, in particular the higher-ranked issues.

Abdul-Aziz Osman (2018) suggests that there should be three levels of modifications in developing countries to overcome the challenges of heritage conservation: strategic, planning and site management. Also, Hegazy (2015) in his research emphasizes the importance of the support and participation of related organizations in developed countries of historical sites in developing societies as human heritage. However, undoubtedly, solving the heritage conservation barriers first helps to preserve the identity of a society (Forster and Kayan, 2009; Kisusi and Masele, 2018), to introduce its heritage to other people and to gain economic benefits (Johar *et al.*, 2011; Kisusi and Masele, 2018) from flourishing cultural tourism, especially in small cities.

Finally, this chapter has some limitations that must be acknowledged. As mentioned in the research methodology, the main data source was the most relevant authorities' interviews about heritage conservation challenges published on the internet. Despite the concentration and efforts of the authors while collecting data, it seems that in-depth and in-person interviews can result in more precise, nuanced results.

Future studies could focus on a different country context or even compare several developing countries. Employing alternative qualitative methodologies can help develop and verify the conceptual model. Also, for a better understanding of the interrelationship between these challenges, more studies and analyses are recommended. In this way, the findings of these studies may lead policy makers and planners to solve these challenges and provide a better set of strategies for heritage conservation and tourism development.

References

Aas, C., Ladkin, A. and Fletcher, J. (2005) Stakeholders' collaboration and heritage management. *Annals of Tourism Research* 32(1), 28–48.

Abdul-Aziz Osman, K. (2018) Heritage conservation management in Egypt: a review of the current and proposed situation to amend it. *Ain Shams Engineering Journal* 9(4), 2907–2916. DOI: 10.1016/j.asej.2018.10.002.

Ahmad, R. (2013) Cultural landscapes as heritage in Malaysia. MA thesis, Utrecht University Repository, Selangor, Malaysia.

Al-Saad, S. (2016) The conflicts between sustainable tourism and urban development in the Jerash archaeological site (Gerasa), Jordan. PhD thesis, The Catholic University of Eichstätt-Ingolstadt, Abu Dhabi, United Arab Emirates.

Al-Tokhais, A. and Thapa, B. (2019) Management issues and challenges of UNESCO World Heritage Sites in Saudi Arabia. *Heritage Tourism* 15(1), 103–110. DOI: 10.1080/1743873X.2019.1594836.

Amar, J.H.N. (2017) Conservation of cultural built heritage: an investigation of stakeholder perceptions in Australia and Tanzania. Ph.D thesis, Bond University, Queensland, Australia.

Amar, J.H.N., Armitage, L. and O'Hare, D. (2016) Built heritage conservation: the implications of cultural racism, segregation and oppression. In: Amoêda, R., Lira, S. and Pinheiro, C. (eds), *The Heritage 2016 – 5th International Conferences on Heritage and Sustainable Development*, Green Lines Institute, pp. 225–236.

Andalib, A. and Abdolahzadefard, A. (2013) Analysis of a development framework for urban heritage conservation versus development trends in Shiraz, Iran. *Architecture and Urban Development* 3(1), 17–23.

Appendino, F. (2017) Balancing heritage conservation and sustainable development: the case of Bordeaux. In: IOP Conference Series: Materials Science and Engineering, Bristol, UK.

Arnstein, S.R. (1969) A ladder of citizen participation. *The American Institute of Planners* 35(4), 216–224. DOI: 10.1080/01944366908977225.

Attride-Stirling, J. (2001) Thematic networks: an analytic tool for qualitative research. *Qualitative Research* 1(3), 385–405. DOI: 10.1177/146879410100100307.

Aziz, A. (2005) Architectural transportation of heritage buildings case study: folk art museum building, Sonargaon, Bangladesh. MA thesis, University of Ottawa, Ontario, Canada.

Aziz, N.Z.M., Razak, A.A., Din, M.A.M. and Nasir, N.M. (2016) Recurring issues in historic building conservation. *Procedia – Social and Behavioral Sciences* 222, 587–595. DOI: 10.1016/j.sbspro.2016.05.217.

Bello, H.E., Bressiani, D., Calliari, T., Marques, M.E.C., Schwengber, E.B. *et al.* (2012) Definition of indicators in the rehabilitation plan of the historic center of Porto Alegre. In: Zancheti, S.M. and Similä, K. (eds) *Measuring Heritage Conservation Performance*. ICCROM, Rome, pp. 244–251.

Boudiaf, B. (2019) The impact of the tourism on the revitalization of the historical center. In: International Conference on Protecting Cultural Heritage of the Muslim World, IRCICA and ISESCO, Istanbul, Turkey.

Braun, V. and Clarke, V. (2006) Using thematic analysis in psychology. *Qualitative Research in Psychology* 3(2), 77–101. DOI: 10.1191/1478088706qp063oa.

Choenni, A. (2015) *Public and Private Roles in Historical Inner-City Revitalization*. Cultural Heritage Agency, Amersfoort, The Netherlands.

CHTM (2020) Iran ranks tenth on UNESCO World Heritage list. Iran's Ministry of Cultural Heritage, Tourism and Handicrafts. Available at: https://www.mcth.ir/english/news/ID/57644 (accessed 21 February 2021).

CHTN (2020) Recording 33000 historic sites in the national heritage list. Available at: https://chtn.ir/news/1399050121/ (accessed 22 February 2021).

Dabrowski, P. (2007) Tourism for conservation, conservation for tourism. Available at: http://www.fao.org/3/v2900e/v2900e09.htm (accessed 18 January 2019).

De La Torre, M. (2013) Values and heritage conservation. *Heritage & Society* 60(2), 155–166. DOI: 10.1179/2159032X13Z.00000000011.

Della Torre, S. (2010) Conservation of built cultural heritage, laws enabling preventive approach: the case of Italy. In: Gustin, M. and Nypan, T. (eds) *Cultural Heritage and Legal Aspects in Europe*. University of Primorska, Koper, Slovenia, pp. 168–178.

Dian, Z.M. and Abdullah, N.C. (2013) Public participation in heritage sites conservation in Malaysia: issues and challenges. *Procedia – Social and Behavioral Sciences* 101, 248–255. DOI: 10.1016/j.sbspro.2013.07.198.

Doganer, S. and Dupont, W. (2013) Cultural heritage tourism and authenticity: San Antonio missions historic district. *WIT Transactions on the Built Environment* 131(13), 15–28. DOI: 10.2495/STR130021.

Domšić, L. (2013) Cultural heritage and identify in the contemporary tourism development. Available at: https://hrcak.srce.hr/file/238105 (accessed 19 November 2019).

Duval, M. and Smith, B. (2013) Rock art tourism in the uKhahlamba/Drakensberg World Heritage Site: obstacles to the development of sustainable tourism. *Sustainable Tourism* 21(1), 134–153. DOI: 10.1080/09669582.2012.699060.

Eboreime, J. (2008) Challenges in heritage conservation in Africa. In: Ndoro, W., Albert, M. and Abungu, G. (eds) *Cultural Heritage and the Law: Protecting Immovable Heritage in English-Speaking Countries in Sub-Saharan Africa*. ICCROM Conservation Studies, Rome, Italy, pp. 1–5.

Farid, S. (2016) Tourism in World Heritage Sites and its impact on economic development: some African countries case studies. In: Doempke, S. (ed.), *Proceedings of the 2nd International Conference on Best Practices in World Heritage: People and Communities*, World Heritage Watch, pp. 721–737.

Fitri, I., Ahmad, Y. and Ratna, S. (2016) Cultural heritage and its legal protection in Indonesia since the Dutch East Indies government period. *Advances in Social Science, Education and Humanities Research* 81, 127–134.

Forster, A.M. and Kayan, B. (2009) Maintenance for historic buildings: a current perspective. *Structural Survey* 27(3), 210–229. DOI: 10.1108/02630800910971347.

Franco, L.M., Newman, J., Murphy, G. and Mariani, M. (1997) *Achieving Quality through Problem Solving and Process Improvement*, 2nd edn. United States Agency for International development (USAID), Washington, DC.

Greg, R. (2018) Cultural tourism: a review of recent research and trends. *Journal of Hospitality and Tourism Management* 36, 12–21. DOI: 10.1016/j.jhtm.2018.03.005.

Günlü, E., Pirnar, E. and Yağcı, K. (2009) Preserving cultural heritage and possible impacts on regional development: case of Izmir. *International Journal of Emerging and Transition Economies* 2(2), 213–229.

Hall, C.M. (2006) Implementing the World Heritage Convention: what happens after listing? In: Leask, A. and Fyall, A. (eds) *Managing World Heritage Sites*. Elsevier, London, pp. 20–34.

Hall, C.M. and McArthur, S. (1998) *Integrated Heritage Management*. Stationery Office, London.

Hall, V.M. (2013) *Handbook of Conservation of Heritage Buildings*. Central Public Works Department, New Delhi, India.

Harun, S.N. (2011) Heritage building conservation in Malaysia: experience and challenges. *Procedia Engineering* 20, 41–53. DOI: 10.1016/j.proeng.2011.11.137.

Hegazy, S.M. (2015) Conservation of historical buildings – the Omani–French museum as a case study. *Housing and Building National Research Center Journal* 11(2), 264–274. DOI: 10.1016/j.hbrcj.2014.03.010.

Henderson, J.C. (2002) Built heritage and colonial cities. *Annals of Tourism Research* 29(1), 254–257. DOI: 10.1016/S0160-7383(01)00009-3.

Imanian, S. (2019) Which of Qazvin's historic sites do you choose to destroy?! Available at: https://www.isna.ir/news/ (accessed 11 January 2020).

Isna News (2017) Was the destruction of Hamedan's Kababian strict legal? Available at: https://www.isna.ir/news/96102513303 (accessed 22 February 2021).

Jamieson, W. (2000) *The Challenges of Sustainable Community Cultural Heritage Tourism*. Asian Institute of Technology, Bangkok, Thailand.

Johar, S., Che-Ani, A.I., Tawil, N.M., Thahir, M.M., Abudullah, N.A.G. *et al.* (2011) Key conservation principles of old traditional mosque in Malaysia. *WSES Transactions on Environment and Development* 4(7), 93–102.

Jokilehto, J. (2017) World heritage and community involvement. In: Nordic World Heritage Conference, Rauma, Norway.

Kammeier, H.D. (2008) Managing cultural and natural heritage resources: part 1 – from concepts to practice. *City and Time* 4(1), 1–13.

Kenny, P. (2006) Developing a best practice model for the re-use and regeneration of vacant buildings at risk in the context of a world heritage site. Bean Conference, John Moores University, Liverpool, UK.

Kisusi, R.L. and Masele, J.J. (2018) Efficacy of public awareness strategies for promoting existing cultural heritage tourism assets in Dar es Salaam. *Journal of Heritage Tourism* 14(2), 117–137. DOI: 10.1080/1743873X.2018.1491980.

Leitao, L. (2011) The protection of world heritage settlements and their surroundings. PhD dissertation, University of Edinburgh, Edinburgh, UK.

Macdonald, S. and Cheong, C. (2014) *The Role of Public–Private Partnerships and the Third Sector in Conserving Heritage Buildings, Sites, and Historic Urban Areas*. Getty Conservation Institute, Los Angeles, California.

Martínez, P.G. (2017) Built heritage conservation and contemporary urban development: the contribution of architectural practice to the challenges of modernisation. *Built Heritage* 1(1), 14–25. DOI: 10.1186/BF03545666.

Martokusumo, W. (2004) Conservation and the modern project: some thoughts on conflict between modernity and tradition. Available at: https://www.academia.edu/9705430/CONSERVATION_AND_THE_MODERN_PROJECT_SOME_THOUGHTS_ON_CONFLICT_BETWEEN_MODERNITY_AND_TRADITION (accessed 18 November 2019).

Matiz, L., Paula, J. and Tonga Uriarte, Y. (2012) Challenges for cultural tourism: conservation and sustainable development. In: *The 2nd Interdisciplinary Tourism Research Conference*.

Mikaeili, M. and Aytuğ, H.K. (2019) Evaluation of Iran's cultural tourism potential from the European Union perspective: Jolfa region. In: Stankov, U., Boemi, S.N., Attia, S., Kostopoulou, S. and Mohareb, N. (eds) *Cultural Sustainable Tourism*. Springer, New York, pp. 115–130.

Millar, S. (1989) Heritage management for heritage tourism. *Tourism Management* 10(1), 9–14. DOI: 10.1016/0261-5177(89)90030-7.

Misra, M. (2009) Revitalisation of Iran's historic buildings: current policies and programs. *International Journal of Environmental Studies* 66(2), 263–270. DOI: 10.1080/00207230902899834.

Mubaideen, S. and Al Kurdi, N. (2017) Heritage conservation and urban development: a supporting management model for the effective incorporation of archaeological sites in the planning process. *Tourism Management* 28, 117–128. DOI: 10.1016/j.culher.2017.05.007.

Munjeri, D. (2004) Anchoring African cultural and natural heritage: the significance of local community awareness in the context of capacity-building. In: de Merode, W., Smeets, R. and Westrik, C. (eds) *Linking Universal and Local Values: Managing a Sustainable Future for World Heritage*. UNESCO, Paris, pp. 75–80.

Ndoro, W. (2008) Legal definitions of heritage. In: Ndoro, W., Mumma, A. and Abungu, G. (eds) *Cultural Heritage and Law*. ICCROM Conservation Studies, Rome, pp. 25–35.

Otieno, D.O. (2015) A protective legislation, policy and practice of archaeological heritage management in Kenya. PhD dissertation, University of Pau and Pays de L'Adour, Pyrénées-Atlantiques, France.

Panteia (2015) Carrying capacity at sensitive cultural heritage sites. Available at: https://eenca.com/eenca/assets/File/EENCA%20publications/Carrying%20capacity%20in%20sensitive%20cultural%20sites.pdf (accessed 23 February 2021).

Patuelli, R., Mussoni, M. and Candela, G. (2013) The effects of World Heritage Sites on domestic tourism: a spatial interaction model for Italy. *Journal of Geographical Systems* 15(3), 369–402. DOI: 10.1007/s10109-013-0184-5.

Penna, K.N., Silvana, C. and Taylor, E. (2013) The challenge of cultural heritage shared management in developing countries. Paper presented at the 8th International Meeting Ciudad Imagen y Memoria, Universidad de Oriente, Santiago de Cuba, Cuba.

Persia Advisor Travel (2017) Iran's heritage in UNESCO. Available at: https://www.persiaadvisor.travel/about-persia/irans-heritages-in-unesco/ (accessed 6 January 2020).

Puren, K. and Jordaan, T. (2014) Towards integrating built heritage resources in urban development through spatial planning. *The Sustainable City* 1, 209–221.

Richards, G. (2018) Cultural tourism: a review of recent research and trends. *Hospitality and Tourism Management* 36, 12–21. DOI: 10.1016/j.jhtm.2018.03.005.

Roque, M.I. and Forte, M.J. (2019) Heritage tourism in Iran, a historical perspective. In: Corria, A., Kozak, M. and Rodrigues, A.I. (eds) *Experiencing Persian Heritage: Perspectives and Challenges*. Emerald Publishing Limited, Bradford, UK, pp. 29–42.

Rouhani, B. (2011) Development and cultural heritage in Iran. ICOMOS 17th General Assembly, Paris, France.

Seyfi, S., Hall, C.M. and Fagnoni, E. (2018) Managing world heritage site stakeholders: a grounded theory paradigm model approach. *Heritage Tourism* 14(4), 308–324. DOI: 10.1080/1743873X.2018.1527340.

Snyder, M.R. (2008) The role of heritage conservation districts in achieving community improvement, UW space. Available at: http://hdl.handle.net/10012/3801 (accessed 22 October 2019).

Spiridon, P. and Sandu, L. (2016) Conservation of cultural heritage: from participation to collaboration. *Enact Journal of Cultural Management and Policy* 5(1), 43–53.

Steinberg, F. (1996) Conservation and rehabilitation of urban heritage in developing countries. *Habitat International* 20(3), 463–475. DOI: 10.1016/0197-3975(96)00012-4.

Throsby, D. (2016) Investment in urban heritage conservation in developing countries: concepts, methods and data. *City, Culture and Society* 7(2), 81–86. DOI: 10.1016/j.ccs.2015.11.002.

Timothy, D.J. and Boyd, S.W. (2015) *Tourism and Trails: Cultural, Ecological and Management Issues*. Channel View Publications, Bristol, UK. DOI: 10.21832/9781845414795.

Tosun, C. (2005) Stages in the emergence of a participatory tourism development approach in the developing world. *Geoforum* 36(3), 333–352. DOI: 10.1016/j.geoforum.2004.06.003.

Tuckett, A.G. (2005) Applying thematic analysis theory to practice: a researcher's experience. *Contemporary Nurse* 19(1–2), 75–87. DOI: 10.5172/conu.19.1-2.75.

Uddhammar, E. (2006) Development, conservation and tourism: conflict or symbiosis? *Review of International Political Economy* 13(4), 656–678. DOI: 10.1080/09692290600839923.

Udeaja, C., Trillo, C., Awuah, K.G.B., Makore, B.C.N., Patel, D.A. *et al*. (2020) Urban heritage conservation and rapid urbanization: insights from Surat, India. *Sustainability* 12(6), 2172. DOI: 10.3390/su12062172.

UNWTO (2020) International tourism highlights. United Nations World Tourism Organization annual report. Available at: https://www.e-unwto.org/doi/epdf/10.18111/9789284422456 (accessed 26 March 2021).

Van Zyl, C.J. (2005) The role of tourism in the conservation of cultural heritage with particular relevance for South Africa. PhD dissertation, University of Stellenbosch, Stellenbosch, South Africa.

Vásquez, M.L.N. (2009) Legal protection of the archeological cultural heritage in Mexico. In: Samantha, W. (ed.) *Proceedings of the 2009 George Wright Society Conference on Parks, Protected Areas, and Cultural Sites*. The George Wright Society, Hancock, Michigan, pp. 227–283.

Walters, T. (2016) Using thematic analysis in tourism research. *Tourism Analysis* 21(1), 107–116. DOI: 10.3727/108354216X14537459509017.

Wang, R., Liu, G., Zhou, J. and Wang, J. (2019) Identifying the critical stakeholders for the sustainable development of architectural heritage of tourism: from the perspective of China. *Sustainability* 11(6), 1671. DOI: 10.3390/su11061671.

Wong, W.H. (2018) Community participation in heritage management: a case study of Hong Kong's conservation approaches. MA thesis, Anhalt University, Bernburg, Germany.

Worthing, D. and Dann, N. (2000) Approaches to the repair of traditional timber-framed buildings: the application of conservation philosophy into practice. *Structural Survey* 18(4), 136–147. DOI: 10.1108/02630800010341462.

Yung, E.H.K. and Chan, E.H.W. (2011) Problem issues of public participation in built-heritage conservation: two controversial cases in Hong Kong. *Habitat International* 35(3), 457–466. DOI: 10.1016/j.habitatint.2010.12.004.

6 Tourism, Conservation and Resident Well-being in Cappadocia (Turkey): An Exploratory Research

Selenay Ata[1] and Sina Kuzuoglu[2*]
[1]Boğaziçi University, Boğaziçi, Turkey; [2]University of Waterloo, Waterloo, Canada

6.1 Introduction

The impacts of increasing tourism activity on residents have a well-established presence in tourism scholarship (Nolan and Seraphin, 2019). Residents' well-being has also become a prominent research avenue on order to more thoroughly understand tourism's impacts and to better inform tourism development initiatives (e.g. Bimonte and D'Agostino, 2020). However, social impacts of tourism development cannot be thought of separately from economic, cultural, environmental and political contexts (Szromek *et al.*, 2019; Ramkissoon, 2020). Considering tourism's influence on and interrelations with numerous attributes of destinations, this chapter focuses on tourism development and regional governance and how they relate to community well-being focusing on the example of the Cappadocia region in central Anatolia, Turkey, a natural and cultural (i.e. mixed) World Heritage Site (WHS).

Tourism activity in the Cappadocia region started in the 1950s. The cultural heritage of Cappadocia alongside its unique natural landscape was instrumental in its inscription as a WHS and is among Turkey's most popular cultural tourism destinations (Tucker and Emge, 2010; Turan and Kozak, 2016). Furthermore, tourism in Cappadocia has been among the first examples of tourism in Turkey to be examined through a sustainability lens (Tosun, 1998). Despite early signs of unbalanced growth and a potentially harmful tourism development trajectory in the long term, tourism in Cappadocia continues to be growth-oriented and competition-based (Çakır *et al.*, 2018) with economic benefits that outweigh its socio-cultural and environmental impacts (Özel and Kozak, 2017; Coban and Yildiz, 2019).

In accordance with the formal partnership between the Applied Tourism Administration and Research Centre at Boğaziçi University (ATARC) and the Planet Happiness project, which aims to collect subjective well-being (SWB) data from WHSs around the globe, this research uses a modified version of the Happiness Index (HI) (Musikanski *et al.*, 2017) that integrates residents' perceptions regarding tourism activity and conservation (Kuzuoglu *et al.*, 2020) as area-specific contextual elements that may influence SWB (Sung and Phillips, 2018). Considering tourism's relatively long presence in the lives of Cappadocian residents and its economic importance for the regional administrative actors, this research considers tourism development and conservation framework as integral components of residents' SWB patterns.

*Corresponding author: skuzuogl@uwaterloo.ca

 Tourism Planning and Development in the Middle East (eds S. Kladou *et al.*)

DOI: 10.1079/9781789249149.0006

From this perspective, this chapter provides an exploratory analysis of SWB components and perceptions of tourism and conservation in Cappadocia, and integrates these findings with the centralized hierarchical heritage governance framework in Turkey (Human, 2015). The aim of this research, thus, is to critically evaluate governance challenges in Cappadocia as a mixed WHS through an evaluation of residents' SWB and to suggest potential pathways for improved, inclusive governance mechanisms. Before moving on to Cappadocia as a WHS and a tourism destination, the next section introduces the literature guiding this chapter.

6.2 Literature Review

This research is centred around the interrelations between tourism activity, the conservation framework and the SWB of Cappadocian residents. To address potential policy interventions that would foster sustainable tourism development that also acknowledges its WHS characteristics, this section first provides an outline of the literature on community and tourism in heritage settings and subsequently introduces SWB in tourism destinations and their related governance issues.

6.2.1 Community, tourism and heritage

The increase in the visibility of tourism's negative social impacts, particularly in destinations experiencing over-tourism (Milano *et al.*, 2019b; Nolan and Seraphin, 2019; Szromek *et al.*, 2019), illustrates the importance of local communities in tourism development. Although tourism is commonly (and, sometimes, mistakenly) revered for its positive economic impacts, it is also associated with environmental problems and socio-cultural transformations. In heritage settings with a tourism component, conservation of heritage assets represents an additional component to consider in the delicate balance between tourism, development and the expectations of local communities (e.g. Wang, 2012).

The historical trajectory of heritage management has resulted in most decision making concerning heritage to be delegated to heritage experts who are responsible for the selection of the heritage to be conserved and for the methods they deem appropriate for conservation (Smith, 2006). The adoption of this perspective has resulted in the deliberate exclusion of community perspective on heritage, which culminated in heritage 'projects [...] that are done *for* communities, rather than *with* them' (Waterton and Smith, 2010, p. 7, emphasis in original). Essentially, the emergence of omnipotent stakeholders in heritage settings as primary decision makers has resulted in a consolidation of power that positioned community at the centre of the narrative but at the periphery of the debate to maintain their privilege (Cortés-Vázquez *et al.*, 2017). In addition to heritage experts at local and national levels, there are also international institutions such as the United Nations Educational, Scientific and Cultural Organization (UNESCO) that oversee the conservation of heritage, specifically of WHSs, which do not necessarily integrate the relationship between heritage and communities into their evaluation process.

Although community engagement is commonly mentioned in management plans (for an analysis of the management plans of natural and mixed WHSs, see Job *et al.*, 2017) and in spite of evidence advocating for local communities' involvement in the management of conservation areas as a contributor to sites' social and environmental integrity (Andrade and Rhodes, 2012), the extent to which communities are actually integrated into the decision-making processes remains a conundrum, particularly in settings with centralized governance structures (Tosun, 1998; Human, 2015). As suggested by Nyaupane and Timothy (2010), residents' interaction with heritage is connected to awareness among local community members, which is instrumental in the conservation of heritage. Both of these factors are positively influenced by education. The importance of the latter is also evident in the study by You *et al.* (2014) of the Wuyishan Scenery District in China, reiterating the importance of locals' active participation in decision making and a fair distribution of benefits and governmental compensations.

Yet it must be acknowledged that prevalence of tourism activity in a region may amplify the difficulty in balancing the needs and expectations of different stakeholders due to the role assigned to tourism in regional development and its direct and indirect impacts on communities' livelihoods, both of which inevitably influence the residents' SWB.

6.2.2 SWB and communities in tourism destinations

SWB can be viewed as a construct that delineates how individuals 'think and experience their lives in positive versus negative ways' (Diener *et al.*, 2018, p. 253) and can be interpreted as 'subjective enjoyment of life-as-a-whole' (Veenhoven, 2012, p. 76). Tourism activity has received increased attention from an SWB perspective with numerous research projects conducted on local communities (e.g. Nawijn and Mitas, 2012; Woo *et al.*, 2015; Bimonte and D'Agostino, 2020; Lindberg *et al.*, 2021).

In areas where tourism has a continuous presence throughout the year, such as in the case of Palma de Mallorca, tourism is suggested to resonate more on residents' SWB (i.e. life satisfaction) as opposed to the affective one (Nawijn and Mitas, 2012). Also, seasonal tourism may influence SWB parameters to varying extents, such as a decline in satisfaction with personal relationships (Bimonte and D'Agostino, 2020). By focusing on tourism's influence on SWB and further tourism development, Woo *et al.* (2015) suggest that both material and non-material components of well-being are positively associated with perceived value of tourism development. Additionally, recent research suggests that the influence of an increase in tourism on SWB is more amplified in areas with higher population density, and is particularly associated with negative environmental impacts (Lindberg *et al.*, 2021). However, perceptions of environmental factors are also influenced by the individuals' attachment and identification with specific places (Kyle *et al.*, 2004).

Alongside the negative socio-economic impacts of tourism, Milano *et al.* (2019a) suggest that 'mainstreaming of special and niche tourism practices in vulnerable places' (p. 355) also contributes to an unbalanced tourism development, which may be connected to tourism planning that is dominated by business and economics perspectives (Moscardo, 2011). Growth-oriented tourism management, primarily emanating from political considerations and market-oriented decision-making processes, may amplify social inequalities that not only negatively influence residents' engagement with tourism but also contribute to a decline in community well-being (Peterson, 2020). Additionally, social comparison with community members may exert a negative influence on social cohesion and resilience in small, close-knit social structures (Rivera *et al.*, 2016). However, the results of Lin *et al.* (2017) suggest that overall life satisfaction has a positive influence on residents' willingness to interact with tourists that culminates in value co-creation. These results underline that the social and cultural benefits of tourism may even be more important than the economic ones. In protected areas, administrative frameworks designed to ensure conservation of heritage resources may also have negative impacts on local communities by influencing their daily lives and traditional practices (García-Frapolli *et al.*, 2009), and also by restricting residents' ownership rights to align them with conservation initiatives (Joy, 2016). Such studies underline the importance of considering socio-cultural and site-specific attributes in tourism management. With this background, the next section introduces the characteristics of Cappadocia as a WHS and a tourism destination, and its governance structure.

6.3 Research Area: Goreme National Park in Cappadocia

Goreme National Park and the rock sites of Cappadocia have been a WHS since 1985 and consist of seven sites (i.e. Goreme National Park, Karain, Karlik, Yesiloz, Soganli villages, and the subterranean cities of Kaymakli and Derinkuyu) in central Anatolia (UNESCO, 1985; see Fig. 6.1). Cappadocia has a unique topography formed through 'differential erosion of the volcanic tuff sediments' (International Union for Conservation of Nature, 1985) that gave rise to fairy chimneys. Throughout history, this topographic peculiarity has been utilized to construct spaces of living, prayer and refuge, which are at the centre of the area's cultural heritage

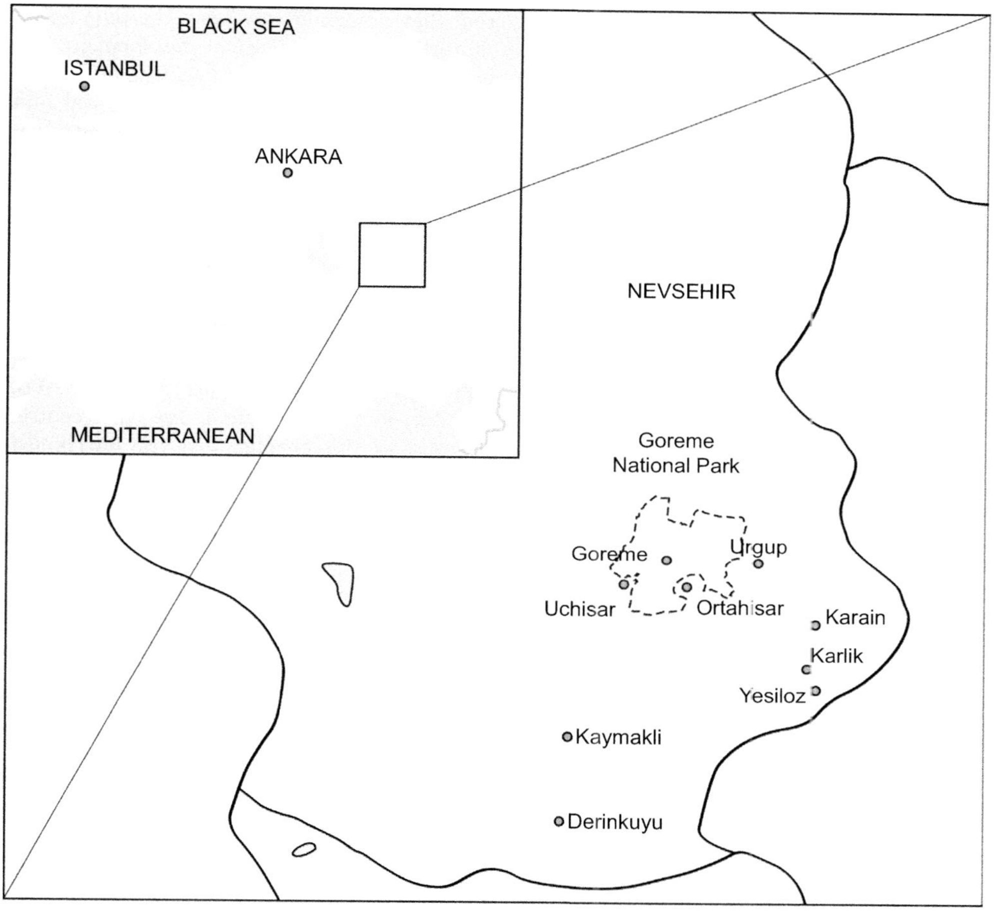

Fig. 6.1. Location and components of Cappadocia WHS in central Anatolia. Source: Gurbuz Publishing (printed with permission).

portfolio and current tourism activity (Çiner and Aydar, 2019).

While tourism had a presence in Cappadocia before (e.g. Turan and Kozak, 2016), it was officially prioritized in the 1980s and the national park became the focal point of tourism in the region (Tucker and Emge, 2010). Goreme National Park is situated on 9574 hectares, with approximately 16,000 residents spread across its 4 districts, i.e. Goreme, Ortahisar, Uchisar and Urgup (TSI, 2020). The remaining components of the WHS can be classified as small villages or hamlets.

According to official accommodation statistics, the annual number of visitors to Cappadocia reached 1.2 million in 2019 and international visitors account for roughly half of this number (Ministry of Culture and Tourism, 2019b). Moreover, 1.4 million people visited the national park in the same year, indicating a significant growth over the preceding 3 years (Ministry of Culture and Tourism, 2020). While this discrepancy indicates considerable daily visits to Cappadocia region, the visitors staying in the national park area also have a high mobility in the broader geographical region.

One of the earliest scholarly inquiries into sustainable tourism development in Turkey focused on Urgup and argued that the limited influence of local stakeholders in decision-making processes constituted the basis for unsustainable tourism development

(Tosun, 1998). The multitude of governmental stakeholders in Cappadocia's management is also noted as a source of potential conflicts in subsequent analyses (Somuncu and Yiğit, 2007; UNESCO, 2014). In multi-stakeholder research in Cappadocia, Tosun (2006) argues that a unified vision in terms of tourism-related decision making does not exist among local stakeholders and the existing top-down approach is primarily rooted in the unbalanced distribution of power among them.

Regardless, tourism has evolved into the most important economic activity in Cappadocia (Tucker, 2010; Coban and Yildiz, 2019). Although tourism's economic benefits may inhibit the recognition of social and environmental problems (Özel and Kozak, 2017), the significant impact of tourism on local residents' livelihoods (Çakır *et al.*, 2018) calls for a careful planning and oversight of tourism that would not only empower local establishments in the tourism scene but would also benefit the local community to a higher extent. However, the local community in Cappadocia is conceptually distant from the heritage resources in the region and prioritizes economic benefits, particularly due to the absence of a planning and coordinating mechanism for tourism development (Yolal and Karacaoglu, 2020). This perspective is also echoed in recent research arguing that establishing a destination management organization in Cappadocia would be beneficial to increase destination competitiveness, to foster sustainability and to address the issues the region faces (Coban and Yildiz, 2019), which are inseparable from heritage conservation and management.

The management of Cappadocia's natural and cultural heritage is tied to various governmental entities due to the multiplicity of conservation frameworks in the area's subregions. However, significant changes to Cappadocia region's governance structure have recently been implemented. First, Cappadocia site management directorate (SMD) was founded in June 2019 (Official Gazette, 2019a) and the national park status of Goreme National Park was revoked thereafter (Official Gazette, 2019b). Although overcoming managerial conflicts to preserve the area's natural characteristics was the official reason for these changes (Ministry of Culture and Tourism, 2019a), critics (e.g. Chamber of Urban Planners, 2019) argue that they predominantly serve the pursuits of tourism growth and urban development and may, in the long term, damage site characteristics. Additionally, Cappadocia's cultural and tourism conservation and development area status was also revoked (Official Gazette, 2020). These administrative changes are essential for the conceptual model guiding this research.

6.4 Conceptual Model

Although the interrelations between SWB, tourism and conservation are the central inquiries of this chapter, they are interpreted from a governance perspective. This approach is particularly relevant for heritage settings in Turkey where the central management structure is commonly fraught with 'highly bureaucratised systems, [...] overall ambiguity on who does what, with consequent problems in terms of responsibilities' (Bonini Baraldi *et al.*, 2013, p. 744). As Morrison *et al.* (2020) argue, although governance issues are rarely addressed by UNESCO, the differential approach to governance in natural and mixed WHSs at the state level may lead to uncontrollable outcomes for sites. In Turkey, centralized governance mechanisms are essential for the selection of the heritage to be conserved, but also result in local communities' deliberate exclusion from heritage management (Human, 2015). In Cappadocia, the recent changes in the governance structure may be interpreted as an attempt to curb the multi-institutional management structure overseeing different aspects of the WHS and to solidify the strict hierarchical bureaucracy reaching from the national to the local level. Della Lucia and Franch (2017) suggest that differences in local context are integral to how different levels of administrators perceive the governance structure surrounding heritage. Furthermore, engagement of local communities in the management process, which supports natural and cultural heritage conservation (Andrade and Rhodes, 2012), is also influenced by the socio-economic attributes of the local context (Della Lucia and Franch, 2017).

In light of the changing governance dynamics and their interconnectedness with social and environmental characteristics, this chapter investigates how residents' perceptions of governmental institutions are associated with their perceptions of environment and cultural heritage conservation, both of which are essential components of tourism activity in the area. The next section outlines the methodology used in this research.

6.5 Methodology

This research is structured as a case study to evaluate the interrelations between SWB and residents' perceptions of tourism and conservation in the specific context of Cappadocia (Yin, 2009) and employs a quantitative approach. In line with the collaboration between ATARC and the Planet Happiness project, which aims to collect SWB data in WHSs across the globe, this research employs the HI in Turkey for the first time, which is briefly described in the next subsection.

6.5.1 Survey instrument

This research uses a modified version of the HI (Musikanski *et al.*, 2017) with additional items concerning residents' perceptions of tourism and conservation that have been localized to the context of Turkey (Kuzuoglu *et al.*, 2020). Essentially, HI is designed to assess the subjective evaluation of various domains of an individual's life (i.e. Cantril ladder; satisfaction with life and affect; psychological well-being; health; time balance; community; social support; lifelong learning, arts and culture; environment; standard of living/economy; and work). With the exception of Cantril ladder and the satisfaction with life and affect domain (response options on 11-point scales), the HI items are rated on 5-point Likert scales (e.g. completely dissatisfied to completely satisfied, completely disagree to completely agree, frequency). There are 50 items in the HI, excluding the demographics. By viewing tourism and conservation as important contextual elements that may influence SWB (Sung and Phillips, 2018), 12 additional items are included on residents' perceptions of tourism (i.e. satisfaction with the number of tourists and willingness for increased tourism (Wang and Pfister, 2008); having financial income from tourism activity (Milman and Pizam, 1988); tourism's impact on quality of life (Faulkner and Tideswell, 1997); tourism's perceived positive influence on conservation (Vareiro *et al.*, 2013); residents' perceptions of conservation framework (i.e. perceived benefit of conservation for future generations; adapted from UNESCO, 1972; satisfaction with conservation efforts, sense of personal responsibility for conservation and awareness of WHS status (You *et al.*, 2014); and conservation efforts' impact on daily life (adapted from Joy, 2016). Length of residence and place of birth are additional demographic items. The next subsection introduces the sampling framework.

6.5.2 Sampling

The sampling design of this research focuses on Goreme National Park, which is not only central to Cappadocia's tourism activity but is also home to the majority of the area's population. In the remainder of this chapter, Cappadocia refers to the Goreme National Park and Cappadocia region refers to the entire WHS. Population statistics in the four districts of Cappadocia (i.e. Goreme, Ortahisar, Uchisar and Urgup) constitute the basis for the sampling. The remaining components of the WHS (i.e. Karlik, Karain and Yesiloz) have a voter population of fewer than 300 and are therefore excluded from the sample.

A convenience sampling approach was used to administer the survey in four districts. The sampling was designed to account for the differences in districts' populations and collected with the aim to be proportionate to these differences. Survey administration was conducted with residents approached randomly on the streets and workplaces located within these districts. The survey was terminated if the respondent did not report as living in one of the districts in the sampling framework. A total of 180 surveys were administered between 15 and 31 March 2019. Two incomplete surveys were eliminated, leaving 178 surveys whose distribution of districts is provided in Table 6.1.

6.5.3 Analysis

Prior to the analysis, Cronbach's alpha values for HI's domains are calculated to test for scale

Table 6.1. Sampling framework.

District	Population	%	Sample	%
Goreme	1,839	11.50	33	18.54
Ortahisar	2,465	15.42	12	6.74
Uchisar	3,215	20.11	26	14.61
Urgup	8,467	52.97	107	60.11
Total	15,986	100.00	178	100.00

reliability. Subsequently, aggregate scores for each domain are computed as the arithmetic mean of the items in respective domains. Due to the exceptionally low means of volunteering and donation patterns (1.41 (SD = 1.01) and 1.84 (SD = 1.24), respectively), these items are not included in the aggregated community domain. Analyses regarding perception of tourism and conservation are conducted on an item basis since they are not structured as pre-tested scales but rather are individually adapted from the relevant literature (Kuzuoglu *et al.*, 2020). In addition to correlation analysis and t-tests to identify the general patterns, a multiple regression model is used to identify the associations between residents' perceptions of governmental institutions and environment, and their satisfaction with conservation using the PROCESS macro (Hayes, 2018).

6.6 Results

The three-step reporting of the results starts with a description of respondents' socio-demographic characteristics and continues with the descriptive statistics of the HI domains and the additional items on tourism and conservation. The last subsection introduces regression results and the mediation analysis.

6.6.1 Socio-demographic profile of respondents

Respondents have a proportionate distribution in terms of their gender and respondents between the ages of 18 and 35 correspond to 54% (n = 96) of the sample. About half of the respondents report being married (49%). In terms of education, 86 respondents (48%) have completed at least one form of tertiary education. Most respondents are aware of Cappadocia's inscription as a WHS (72%) and do not report having a financial income from tourism (72%). Table 6.2 provides an outline of respondents' socio-demographic characteristics.

6.6.2 Descriptive statistics

When one observes the general characteristics of the results of the HI, most of the domains have relatively high average scores for the respondents. In particular, psychological well-being and social support mechanisms are perceived most favourably. However, respondents' perceived quality of time balance and their perceptions regarding the governmental institutions is lower compared to the rest of the HI domains. Table 6.3 illustrates the items in the survey, the descriptive statistics of the domains, as well as the Cronbach's alpha reliability statistics.

The results for additional variables on residents' perceptions of tourism and conservation suggest that the respondents generally have positive views towards tourism development, which they perceive to be beneficial for their quality of life. As Table 6.4 illustrates, cultural heritage conservation is important for the respondents, who, however, are relatively dissatisfied with conservation efforts in Cappadocia.

Additionally, the associations between the demographic variables and the focal variables of this research (i.e. perception of government, environment and satisfaction with conservation) are investigated through bivariate correlations and *t*-tests. Individuals with a higher education have an increased awareness of WHS status ($t(176) = 5.18$, $p < 0.001$) but lower perceptions of government ($t(173) = -3.27$, $p < 0.01$) and satisfaction with conservation ($t(176)$

Table 6.2. Socio-demographic profile of respondents.

Variable	Category	n	%
Gender	Female	87	48.88
	Male	91	51.12
Marital status	Without partner (single, divorced, widowed)	91	51.12
	With partner (married)	87	48.88
Work	Employed	109	61.24
	Not employed	69	38.76
Education	Basic education (high school or less)	92	51.68
	Higher education (some tertiary degree)	86	48.32
Tourism income	No	128	71.91
	Yes	50	28.09
WHS awareness	No	49	27.53
	Yes	129	72.47
Age (Mean = 37.54 SD = 15.82)	18 to 25	54	30.34
	26 to 35	42	23.60
	36 to 45	30	18.85
	46 to 55	24	13.48
	56 to 65	14	7.87
	66 and above	14	7.87

= -3.19, $p < 0.01$). Perceptions of environment positively correlate with age ($r(176) = 0.41$, $p < 0.001$) and with satisfaction with conservation ($r(176) = 0.23$, $p < 0.01$) and are higher among respondents with a partner ($t(176) = 5.71$, $p < 0.001$). Respondents' perceptions of government have a positive association with their perceptions of environment ($r(173) = 0.26$, $p < 0.001$) and their satisfaction with conservation ($r(173) = 0.44$, $p < 0.001$). The next subsection introduces the results of the regression model.

6.6.3 Multiple regression results

The multiple regression model used in this research has perceptions of government as the dependent, and perceptions of environment and satisfaction with conservation as the independent variables. Gender, age, education, marital status, tourism income and WHS awareness are the control variables in the model.

As may be observed in the regression results in Table 6.5, the coefficient of residents' perceptions of the environment in Model 2 decreases with the addition of satisfaction with conservation from 0.32 to 0.21, and the adjusted R^2 improves. Additionally, in Model 3, the statistical significance of the education variable is also diminished. The simple mediation analysis using PROCESS macro (Hayes, 2018) with 5000 bootstrap samples suggests that satisfaction with conservation acts as a partial mediator between residents' perceptions of environment and their perceptions regarding governmental institutions (see Table 6.6 and Fig. 6.2).

6.7 Discussion

The Cappadocia region has evolved into an important mature cultural and natural destination in Turkey with a considerable reliance on tourism for local livelihood (Coban and Yildiz,

Table 6.3. Descriptive statistics of HI domains.

Domain	Survey item	n	Mean	SD	Alpha
Cantril ladder[a]		178	6.62	2.30	-
Satisfaction with life and affect	Overall satisfaction with life[a] Feeling of doing worthwhile things in life[a] Level of happiness the day before[a] Level of anxiety the day before[a, b]	178	6.64	1.91	0.73
Psychological well-being	Having a purposeful and meaningful life Engagement with and interest in daily activities Optimism about one's future Sense of accomplishment from what one does Having positive feelings about oneself	178	4.06	0.67	0.72
Health	Perceived level of health Having a lot of energy the week before Satisfaction with one's ability to perform daily living activities Satisfaction with the one's quality of exercise	163	3.41	0.76	0.55
Time balance	Amount of time spent on doing the things that one enjoys Having a rushed life[b] Having plenty of spare time	178	2.75	0.98	0.75
Community	Sense of belonging to the local community Number of neighbours one trusts Number of local businesses one trusts Perceived likelihood of a lost wallet being returned by someone living close by Satisfaction with personal safety	178	3.07	0.89	0.77
Social support	Satisfaction with personal relationships People in my life care about me The amount of time one felt loved the week before The amount of time one felt lonely the week before[b]	177	3.87	0.79	0.75
Lifelong learning, arts and culture	Satisfaction with access to sports and recreational activities Satisfaction with access to artistic and cultural activities Satisfaction with access to activities to develop skills by informal education The frequency of feeling uncomfortable/out of place in one's neighbourhood[b]	178	3.52	0.96	0.67
Environment	Perceived healthiness of one's physical environment Satisfaction with efforts to preserve natural environment Satisfaction with the opportunities to enjoy nature Satisfaction with air quality	178	3.62	0.86	0.75

Continued

Table 6.3. Continued

Domain	Survey item	n	Mean	SD	Alpha
Government	Perceived widespread corruption in local government[b] Public officials in one's city or town pay attention to what people think Level of confidence in national government Level of confidence in local government	175	2.74	1.08	0.81
Standard of living/economy	Level of stress about personal finances Frequency of living pay cheque to pay cheque Frequency of eating less because there wasn't enough food or money Having enough money to buy things one wants	177	3.58	0.99	0.68
Work	Satisfaction with current work life Satisfaction with balance between time spent on job and other aspects of life The amount of time one finds current work life interesting Perception of appropriate pay in job Perceived level of productivity at work Ability to decide on how one does their job	136	3.13	0.81	0.79

[a]11-point scale
[b]reverse-coded

Table 6.4. Descriptive statistics of tourism- and conservation-related items.

Domain	Survey item	Mean	SD
Tourism	Satisfaction with the number of tourists	3.19	1.20
	Willingness to accept more tourists	4.45	0.97
	Tourism's perceived positive impact on quality of life	4.01	1.14
	Positive effect of tourism prospects on conservation and restoration of heritage assets	3.80	1.23
Conservation	Importance of heritage conservation for future generations	4.37	1.00
	Satisfaction with conservation efforts	2.94	1.38
	Sense of personal responsibility for conservation of heritage assets	3.62	1.21
	Conservation efforts' positive effects on daily life	3.22	1.10

n = 178

2019). Although tourism in Cappadocia and its sustainable development prospects have become subjects of scholarly inquiry fairly early (Tosun, 1998), tourism development mechanisms constitute most of the subsequent research (e.g. Turan and Kozak, 2016; Coban and Yildiz, 2019). This research deviates from this by positioning residents' subjective perceptions about different aspects of their lives to assess tourism and conservation in Cappadocia from a governance perspective.

The unique natural landscape of the Cappadocia region and the human influence that transformed these natural characteristics to heritage assets (Çiner and Aydar, 2019) are central to the tourism activity taking place therein. Although environmental issues have been raised as an outcome of tourism activity

Table 6.5. Unstandardized coefficients for regression models for residents' government perception.

	Model 1		Model 2		Model 3	
Variables	B		B		B	
Gender	0.24		0.28		0.31	*
Age	-0.00		-0.01		-0.01	
Education	-0.49	**	-0.40	*	-0.16	
Marital status	0.50	*	0.36		0.41	*
Tourism income	-0.03		-0.06		0.01	
WHS awareness	0.35		0.50	*	0.58	**
Environment	-		0.32	**	0.21	*
Conservation satisfaction	-		-		0.31	***
Adjusted R^2	0.10		0.15		0.29	

n = 175, * $p < 0.05$, ** $p < 0.01$, *** $p < 0.001$

Table 6.6. Bootstrap analysis of total effect, direct effect and indirect effects.

			Bootstrapping	
			Confidence interval (CI)	
	Effect	SE	Lower limit 95% CI	Upper limit 95% CI
Total effect (c)	0.3205	0.1027	0.1177	0.5234
Direct effect (c')	0.2098	0.0956	0.0210	0.3985
Indirect effect	0.1108	0.0524	0.0209	0.2211

Focal variable = Perceptions of environment
Criterion variable = Perceptions of government
Mediator = Satisfaction with conservation

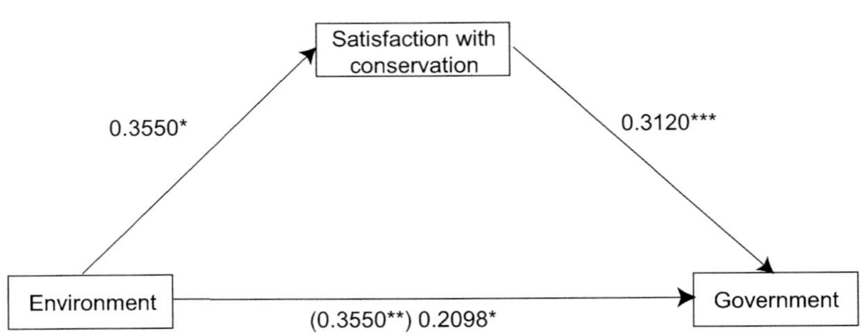

Fig. 6.2. Model of total effect, direct effect and indirect effect for residents' government perceptions. Note: *$p < 0.05$, **$p < 0.01$, ***$p < 0.001$.

(Özel and Kozak, 2017) and of developmentalist perspectives (Chamber of Urban Planners, 2019) in the region, the results of this research suggest that residents have a generally favourable view of environmental characteristics in the region, which increases with age and is positively correlated with marital status, as is residents' satisfaction with conservation. Thus, the connection between residents and their social circles may be interpreted as factors influencing their perceptions of environmental characteristics (Kyle *et al.*, 2004). Results also suggest education to be an important factor influencing residents' perceptions of cultural heritage, supporting previous literature (Nyaupane and Timothy, 2010; You *et al.*, 2014).

Although community engagement in management is associated with improved conservation of natural and cultural assets (Andrade and Rhodes, 2012), the centralized and strict hierarchical governance system in Turkey systematically excludes residents (Bonini Baraldi *et al.*, 2013; Human, 2015). Essentially, the findings linking environmental issues and cultural heritage-related decision making with the effectiveness of governmental institutions in Cappadocia may be interpreted as the local community's detachment from conservation frameworks, which may culminate in the deterioration of natural and cultural heritage assets that are essential tourism resources (Andrade and Rhodes, 2012). Thus, Cappadocia's growth-oriented tourism strategy (Tucker and Emge, 2010), through its intentional distance from local stakeholders including residents (Moscardo, 2011), may not only have negative consequences for Cappadocia's tourism prospects but may also result in diminishing community well-being (Rivera *et al.*, 2016; Peterson, 2020).

The changes in the governance structure of Cappadocia (Official Gazette, 2019a) may be interpreted as the intentional creation of a single powerful administrative stakeholder in Cappadocia region that acts a proxy for the central government (Cortés-Vázquez *et al.*, 2017). Although achieving more efficient decision making through a simplified administrative framework represents the official rationale for the creation of the SMD (Ministry of Culture and Tourism, 2019a), essentially, it is an extension of the hierarchical bureaucratic system established throughout its modern history. Considering the relatively negative perception of governmental institutions among Cappadocian residents and its association with (natural and cultural) conservation-related issues, the next section introduces research implications, as well as its limitations and directions for future research.

6.8 Implications, Limitations and Future Research Directions

This research has started with the assumption that tourism and conservation are two inseparable aspects of residents' lives in Cappadocia. The SWB perspective employed here provides an alternative framework to evaluate the impacts of tourism development and conservation in natural and cultural protected sites. Drawing on the growing attention in the scholarly literature to local communities, and their well-being in tourism destinations and heritage settings (e.g. Waterton and Smith, 2010; Bimonte and D'Agostino, 2020; Lindberg *et al.*, 2021), this research focused on the association between how residents perceive environmental and cultural conservation, and their views regarding the governance aspects of their lives. Two major theoretical implications emanate from our findings: First, the integration of residents' perceptions of government with WHS management framework suggests the possibility of contextualizing residents based on governance attributes of specific sites. Second, our findings on the relationship between environmental perceptions and cultural heritage conservation suggest that for the residents of mixed WHSs, natural and cultural conservation may be intertwined, highlighting the importance of a holistic approach to conservation management in such areas.

In Turkey, the central management structure regarding heritage and tourism has limited local communities' involvement in decision-making processes surrounding their cities (Human, 2015), and, thus, the associations introduced in this research are examined in such a scenario. Although environmental concerns are particularly relevant for Cappadocia due to its natural attributes and the signs of negative environmental impacts through tourism (Özel and Kozak, 2017), these are not reflected in the findings. However, the association of environmental perception with the demographic variables of age and marital status, both of which are conceptually associated with sense of belonging and attachment, may indicate the existence of additional latent constructs that may influence different SWB domains (Ramkissoon, 2020). Considering the recent changes in Cappadocia's governance structure, the findings of this research have important implications for the newly founded SMD.

First, the consolidation of power in a single governmental entity may further diminish the stakeholder involvement in decision making (Cortés-Vázquez *et al.*, 2017), which is not confined to residents but also includes (local) organizations with expertise and know-how in the Cappadocia region. An inclusive and transparent organizational structure of the SMD, informed by local stakeholders, would not only contribute to community well-being and heritage conservation but would also support residents' engagement with conservation management and tourism development. This would also help to alleviate the tensions between residents and governmental institutions. Second, given that the Cappadocia WHS is dispersed over a large geographical area, differences at the local level may jeopardize effective implementation of decisions made at regional and national levels (Della Lucia and Franch, 2017). Thus, the SMD would benefit from integrating local administrative challenges and opportunities as well as socio-economic discrepancies into its strategic framework. Finally, the SMD may assume the role of the coordinating and monitoring mechanism for tourism development suggested by Yolal and Karacaoglu (2020) in order to minimize negative tourism impacts on local community and to ensure compliance with natural and cultural conservation framework(s).

Although the Cappadocia region has previously experienced external shocks to its tourism activity through regional and national issues (e.g. increased threat of terror attacks around 2015), the COVID-19 pandemic that started in early 2020 has resulted in plummeting visitor numbers in Cappadocia and over-representation of daily trips to the region. Furthermore, the pandemic-related restrictions severely limited the revenue-generating tourism resources (e.g. accommodation facilities, hot-air balloons). Although diverse economic structures may act as a buffer for the livelihood of local communities, a recent report by the Association of Turkish Travel Agencies (2020) continues to advocate for growth-oriented tourism marketing and promotion strategies in support of the Cappadocia region's recovery with limited interaction with local stakeholders. In accordance with the findings of this research and the experiences the region accumulated through the pandemic, it is suggested that the perspectives of local stakeholders, including residents, should be incorporated into the area's recovery strategy, which does not solely focus on reviving tourism but also on the region's broader sustainability objectives.

The findings of this research and its implications, however, are subject to some limitations. First, the convenience sampling process and the small sample may not reflect Cappadocia's population characteristics, particularly when one considers the respondents' relatively high education levels. Additionally, the data collection was conducted prior to local elections, which may have negatively influenced researchers' access to residents in different districts, leading to under- and over-representation of the districts in the sample. Second, this research uses a localized version of the HI, some items of which may not be culturally compatible with Turkey. Third, due to its case study nature, the findings of this research may not apply to different mixed WHSs in Turkey, or to settings with different socio-cultural and environmental characteristics.

In line with these limitations future research in the Cappadocia region would benefit from alternative data collection strategies with alterations in the sampling process and also including qualitative and mixed-method insights to gain a deeper understanding of factors influencing residents' SWB. Furthermore, a longitudinal data collection is suggested to identify the causal relationships between tourism, conservation and SWB patterns as well as potential seasonality effects. In the case of the Cappadocia region, investigating a broader tourism-influenced geographical area would provide a more thorough assessment of tourism and conservations' impacts on residents' SWB. In addition to these Cappadocia-specific potential research avenues, investigations of SWB in relatively underdeveloped mixed WHS destinations would contribute to a more detailed account of tourism and conservation's impacts on SWB and how they are influenced by tourism development level. Finally, this research has focused on an area where the natural and cultural landscape is an integral part of residents' lives and is predominantly valorized through tourism. Future research would also benefit from focusing on regions in which residents are connected through different mechanisms to their natural (e.g. natural resource use, agriculture) and cultural landscapes (e.g. functional use of cultural heritage assets).

Acknowledgements

The authors extend their gratitude to Bengi Ertuna for her gracious support and guidance throughout this research. Furthermore, we would like to thank Burçin Hatipoğlu for her contributions in various steps of this research, and the editors for their comments and suggestions on an earlier version of this chapter.

References

Andrade, G.S.M. and Rhodes, J.R. (2012) Protected areas and local communities: an inevitable partnership toward successful conservation strategies? *Ecology and Society* 17(4), 14. DOI: 10.5751/ES-05216-170414.

Association of Turkish Travel Agencies (2020) COVID-19'un turizme etkileri ve Nevşehir turizminin geleceği. [Impacts of COVID-19 on tourism and the future of Nevsehir tourism]. Available at: https://www.tursab.org.tr/assets/assets/uploads/ticaret-odasi/nevsehir-r.pdf (accessed 20 July 2021).

Bimonte, S. and D'Agostino, A. (2020) Tourism development and residents' well-being: comparing two seaside destinations in Italy. *Tourism Economics* 27(7), 1508–1525. DOI: 10.1177/1354816620916962.

Bonini Baraldi, S., Shoup, D. and Zan, L. (2013) Understanding cultural heritage in Turkey: institutional context and organisational issues. *International Journal of Heritage Studies* 19(7), 728–748. DOI: 10.1080/13527258.2012.700283.

Chamber of Urban Planners (2019) Göreme milli park statüsü korunarak yönetilmelidir! [Goreme should be managed by preserving its national park status]. Available at: http://www.tmmob.org.tr/en/node/15188 (accessed 15 April 2021).

Coban, G. and Yildiz, O.S. (2019) Developing a destination management model: case of Cappadocia. *Tourism Management Perspectives* 30, 117–128. DOI: 10.1016/j.tmp.2019.02.012.

Cortés-Vázquez, J., Jiménez-Esquinas, G. and Sánchez-Carretero, C. (2017) Heritage and participatory governance: an analysis of political strategies and social fractures in Spain. *Anthropology Today* 33(1), 15–18. DOI: 10.1111/1467-8322.12324.

Çakır, O., Evren, S., Tören, E. and Kozak, N. (2018) Utilizing the sustainable livelihoods approach to evaluate tourism development from the rural host communities' point of view: the case of Cappadocia (Turkey). *GeoJournal of Tourism and Geosites* 21, 7–25.

Çiner, A. and Aydar, E. (2019) A fascinating gift from volcanoes: the fairy chimneys and underground cities of Cappadocia. In: Kuzucuoglu, C., Çiner, A. and Kazancı, N. (eds) *Landscapes and Landforms of Turkey*. Springer, Berlin, pp. 535–549. DOI: 10.1007/978-3-030-03515-0.

Della Lucia, M. and Franch, M. (2017) The effects of local context on World Heritage Site management: The Dolomites Natural World Heritage Site, Italy. *Journal of Sustainable Tourism* 25(12), 1756–1775. DOI: 10.1080/09669582.2017.1316727.

Diener, E., Oishi, S. and Tay, L. (2018) Advances in subjective well-being research. *Nature Human Behaviour* 2(4), 253–260. DOI: 10.1038/s41562-018-0307-6.

Faulkner, B. and Tideswell, C. (1997) A framework for monitoring community impacts of tourism. *Journal of Sustainable Tourism* 5(1), 3–28. DOI: 10.1080/09669589708667273.

García-Frapolli, E., Ramos-Fernández, G., Galicia, E. and Serrano, A. (2009) The complex reality of biodiversity conservation through natural protected area policy: three cases from the Yucatan Peninsula, Mexico. *Land Use Policy* 26(3), 715–722. DOI: 10.1016/j.landusepol.2008.09.008.

Hayes, A.F. (2018) *Introduction to Mediation, Moderation, and Conditional Process Analysis: A Regression-Based Approach*, 2nd edn. Guilford Press, New York.

Human, H. (2015) Democratising world heritage: the policies and practices of community involvement in Turkey. *Journal of Social Archaeology* 15(2), 160–183. DOI: 10.1177/1469605314566557.

International Union for Conservation of Nature (1985) Nomination to the World Heritage List – Goreme Valley. Available at: https://whc.unesco.org/document/153261 (accessed 30 March 2021).

Job, H., Becken, S. and Lane, B. (2017) Protected areas in a neoliberal world and the role of tourism in supporting conservation and sustainable development: an assessment of strategic planning, zoning, impact monitoring, and tourism management at natural World Heritage Sites. *Journal of Sustainable Tourism* 25(12), 1697–1718. DOI: 10.1080/09669582.2017.1377432.

Joy, C. (2016) Enchanting town of mud': Djenné, a World Heritage Site in Mali. In: De Jong, F. and Rowlands, M. (eds) *Reclaiming Heritage: Alternative Imaginaries of Memory in West Africa*. Routledge, London, UK, pp. 145–159. DOI: 10.4324/9781315421131-6.

Kuzuoglu, S., Ata, S., Hatipoglu, B. and Ertuna, B. (2020) Subjective well-being in World Heritage Sites: localization and thematic integration of the happiness index for Turkey. *International Journal of Community Well-Being* 3(2), 223–240. DOI: 10.1007/s42413-020-00057-8.

Kyle, G., Graefe, A., Manning, R. and Bacon, J. (2004) Effects of place attachment on users' perceptions of social and environmental conditions in a natural setting. *Journal of Environmental Psychology* 24(2), 213–225. DOI: 10.1016/j.jenvp.2003.12.006.

Lin, Z., Chen, Y. and Filieri, R. (2017) Resident-tourist value co-creation: the role of residents' perceived tourism impacts and life satisfaction. *Tourism Management* 61, 436–442. DOI: 10.1016/j.tourman.2017.02.013.

Lindberg, K., Munanura, I.E., Kooistra, C., Needham, M.D. and Ghahramani, L. (2021) Understanding effects of tourism on residents: a contingent subjective well-being approach. *Journal of Travel Research* 61(2), 346–361. DOI: 10.1177/0047287520988912.

Milano, C., Novelli, M. and Cheer, J.C. (2019a) Overtourism and tourismphobia: a journey through four decades of tourism development, planning and local concerns. *Tourism Planning & Development* 16(4), 353. DOI: 10.1080/21568316.2019.1599604.

Milano, C., Novelli, M. and Cheer, J.M. (2019b) Overtourism and degrowth: a social movements perspective. *Journal of Sustainable Tourism* 27(12), 1857–1875. DOI: 10.1080/09669582.2019.1650054.

Milman, A. and Pizam, A. (1988) Social impacts of tourism on central Florida. *Annals of Tourism Research* 15(2), 191–204. DOI: 10.1016/0160-7383(88)90082-5.

Ministry of Culture and Tourism (2019a) Kamuoyuna [Public announcement]. Available at: https://basin.ktb.gov.tr/TR-246153/kamuoyuna.html (accessed 15 April 2021).

Ministry of Culture and Tourism (2019b) Konaklama istatistikleri [Accommodation statistics]. Available at: https://yigm.ktb.gov.tr/TR-201120/konaklama-istatistikleri.html (accessed 10 January 2020).

Ministry of Culture and Tourism (2020) Müze istatistikleri [Museum statistics]. Available at: http://www.dosim.gov.tr/muze-istatistikleri (accessed 21 July 2021).

Morrison, T.H., Adger, W.N., Brown, K., Hettiarachchi, M., Huchery, C. *et al.* (2020) Political dynamics and governance of World Heritage Ecosystems. *Nature Sustainability* 3(11), 947–955. DOI: 10.1038/s41893-020-0568-8.

Moscardo, G. (2011) Exploring social representations of tourism planning: issues for governance. *Journal of Sustainable Tourism* 19(4–5), 423–436. DOI: 10.1080/09669582.2011.558625.

Musikanski, L., Cloutier, S., Bejarano, E., Briggs, D., Colbert, J. *et al.* (2017) Happiness index methodology. *Journal of Social Change* 9(1), 4–31. DOI: 10.5590/JOSC.2017.09.1.02.

Nawijn, J. and Mitas, O. (2012) Resident attitudes to tourism and their effect on subjective well-being. *Journal of Travel Research* 51(5), 531–541. DOI: 10.1177/0047287511426482.

Nolan, E. and Seraphin, H. (2019) Venice: capacity and tourism. In: Dodds, R. and Butler, R.W. (eds) *Overtourism: Issues, Realities and Solutions*. De Gruyter, Berlin, pp. 139–151. DOI: 10.1515/9783110607369-010.

Nyaupane, G.P. and Timothy, D.J. (2010) Heritage awareness and appreciation among community residents: perspectives from Arizona, USA. *International Journal of Heritage Studies* 16(3), 225–239. DOI: 10.1080/13527251003620776.

Official Gazette (2019a) Law nr. 7174 – Kapadokya alanı hakkında kanun [Law regarding Cappadocia site]. Available at: https://www.resmigazete.gov.tr/eskiler/2019/06/20190601-11.htm (accessed 10 April 2021).

Official Gazette (2019b) Presidential decree nr. 1673. Available at: https://www.resmigazete.gov.tr/eskiler/2019/10/20191022-3.pdf (accessed 10 April 2021).

Official Gazette (2020) Presidential decree nr. 2155. Available at: https://www.resmigazete.gov.tr/eskiler/2020/02/20200225-29.pdf (accessed 10 April 2021).

Özel, Ç.H. and Kozak, N. (2017) An exploratory study of resident perceptions toward the tourism industry in Cappadocia: a social exchange theory approach. *Asia Pacific Journal of Tourism Research* 22(3), 284–300. DOI: 10.1080/10941665.2016.1236826.

Peterson, R.R. (2020) Over the Caribbean top: community well-being and over-tourism in small island tourism economies. *International Journal of Community Well-Being* 1–38. DOI: 10.1007/s42413-020-00094-3.

Ramkissoon, H. (2020) Perceived social impacts of tourism and quality-of-life: a new conceptual model. *Journal of Sustainable Tourism* 1–17. DOI: 10.1080/09669582.2020.1858091.

Rivera, M., Croes, R. and Lee, S.H. (2016) Tourism development and happiness: a residents' perspective. *Journal of Destination Marketing & Management* 5(1), 5–15. DOI: 10.1016/j.jdmm.2015.04.002.

Smith, L. (2006) *Uses of Heritage*. Routledge, New York. DOI: 10.4324/9780203602263.

Somuncu, M. and Yiğit, T. (2007) "Türkiye'deki dünya mirası alanlarının Güncel durumlarının saptanması raporu: Göreme Milli Parkı ve Kapadokya kayalık sitleri [Report on determining the current status of World Heritage Sites in Turkey: Göreme National Park and the rock sites of Cappadocia]." Ankara, Turkey: UNESCO Turkey National Commission.

Sung, H. and Phillips, R.G. (2018) Indicators and community well-being: exploring a relational framework. *International Journal of Community Well-Being* 1(1), 63–79. DOI: 10.1007/s42413-018-0006-0.

Szromek, A.R., Kruczek, Z. and Walas, B. (2019) The attitude of tourist destination residents towards the effects of overtourism – Kraków case study. *Sustainability* 12(1), 228. DOI: 10.3390/su12010228.

Tosun, C. (1998) Roots of unsustainable tourism development at the local level: the case of Urgup in Turkey. *Tourism Management* 19(6), 595–610. DOI: 10.1016/S0261-5177(98)00068-5.

Tosun, C. (2006) Expected nature of community participation in tourism development. *Tourism Management* 27(3), 493–504. DOI: 10.1016/j.tourman.2004.12.004.

TSI (2020) Adrese dayalı nüfus kayıt sistemi sonuçları [Address-based population registry system results]. Turkish Statistical Institute. Ankara, Turkey. Available at: https://biruni.tuik.gov.tr/medas/?kn=95&locale=tr (accessed 9 March 2021).

Tucker, H. (2010) Peasant-entrepreneurs. *Annals of Tourism Research* 37(4), 927–946. DOI: 10.1016/j.annals.2010.03.001.

Tucker, H. and Emge, A. (2010) Managing a World Heritage Site: the case of Cappadocia. *Anatolia – An International Journal of Tourism and Hospitality Research* 21(1), 41–54. DOI: 10.1080/13032917.2010.9687089.

Turan, A. and Kozak, N. (2016) Residents' support for tourism in Cappadocia: case of the hippies. *The Social Sciences* 11(20), 4875–4881.

UNESCO (1972) Convention concerning the protection of the world cultural and natural heritage. Available at: https://whc.unesco.org/en/conventiontext/ (accessed 30 March 2021).

UNESCO (1985) Göreme National Park and the rock sites of Cappadocia. Available at: https://whc.unesco.org/en/list/357/ (accessed 30 March 2021).

UNESCO (2014) Periodic reporting cycle 2 – section II – Göreme National Park and the rock sites of Cappadocia. Available at: https://whc.unesco.org/documents/164430 (accessed 10 January 2020).

Vareiro, L.M. da C., Remoaldo, P.C. and Cadima Ribeiro, J.A. (2013) Residents' perceptions of tourism impacts in Guimarães (Portugal): a cluster analysis. *Current Issues in Tourism* 16(6), 535–551. DOI: 10.1080/13683500.2012.707175.

Veenhoven, R. (2012) Happiness: also known as 'life satisfaction' and 'subjective well-being. In: Land, K.C., Michalos, A.C. and Sirgy, M.J. (eds) *Handbook of Social Indicators and Quality of Life Research*. Springer, Dordrecht, The Netherlands, pp. 63–77. DOI: 10.1007/978-94-007-2421-1_3.

Wang, S.-Y. (2012) From a living city to a World Heritage City: authorised heritage conservation and development and its impact on the local community. *International Development Planning Review* 34(1), 1–17. DOI: 10.3828/idpr.2012.2.

Wang, Y. and Pfister, R.E. (2008) Residents' attitudes toward tourism and perceived personal benefits in a rural community. *Journal of Travel Research* 47(1), 84–93. DOI: 10.1177/0047287507312402.

Waterton, E. and Smith, L. (2010) The recognition and misrecognition of community heritage. *International Journal of Heritage Studies* 16(1–2), 4–15. DOI: 10.1080/13527250903441671.

Woo, E., Kim, H. and Uysal, M. (2015) Life satisfaction and support for tourism development. *Annals of Tourism Research* 50, 84–97. DOI: 10.1016/j.annals.2014.11.001.

Yin, R. (2009) *Case Study Research: Design and Methods*. Sage, Thousand Oaks, California.

Yolal, M. and Karacaoglu, S. (2020) An evaluation of tourism potential and sustainability in Cappadocia, Turkey. In: Cravidai, F. and Santos, N. (eds) *Management of World Heritage Sites: Cultural Landscapes and Sustainability*. Cambridge Scholars Publishing, Newcastle upon Tyne, UK, pp. 73–91.

You, W., He, D., Hong, W., Liu, C., Wu, L. *et al*. (2014) Local people's perceptions of participating in conservation in a heritage site: a case study of the Wuyishan Scenery District cultural and natural heritage site in southeastern China. *Natural Resources Forum* 38(4), 296–307. DOI: 10.1111/1477-8947.12057.

7 Moving Away from Wine Tourism and Sustainable Tourism Development: The Paradox in the Case of Bozcaada (Tenedos), Turkey

Nurşah Ayhan[1]* and Stella Kladou[2]
[1]University Carlos III of Madrid, Getafe, Madrid, Spain; [2]Hellenic Mediterranean University, Heraklion, Crete, Greece

7.1 Introduction

Sustainable tourism development and place branding are two, perhaps equally popular, concepts in the tourism literature and practice. Sustainable development is a process meeting the audiences' needs at a socially balanced level, protecting environmental values and enforcing economic growth, whereas place branding is a developmental and participatory action (Mitropoulou and Spilanis, 2020; Rebelo *et al.*, 2020). Place branding needs to essentially integrate sustainable tourism development and represent 'the nature of a place's identity' (Campelo *et al.*, 2014). However, scholars tend to treat them independently, which contradicts the fact that the primary success of sustainable tourism development and branding efforts lies behind how a destination is defined in a place. 'Destination' is closely attached to a place, supported by branding efforts revealing and differentiating its characteristics to eventually reach a system of experiences and communications to be marketed (Blain *et al.*, 2005; Hultman and Hall, 2012; Campelo *et al.*, 2014). Destination branding is 'a network of associations in consumers' minds' (Zenker and Braun, 2010, p. 1). Thus, considering the overall design, different means of expression of the destination, values, culture and the relationships among stakeholders is vital. Although stakeholders often mistakenly consider branding as a promotion tool, branding, the identity of the place and inclusive stakeholders' involvement have to be strategically orchestrated to pursue sustainable development (Zouganeli *et al.*, 2012). The place brand should be a 'relationship-builder' concerning the locality of the place, instead of a 'promotional vehicle', whereas branding should enable the place identity (Giovanardi *et al.*, 2013; Kavaratzis and Hatch, 2013).

Along these lines, wine emerges as an agricultural product that is intertwined with the place, local culture and identity (Thode and Maskulka, 1998). Destinations exist with one-of-a-kind identity, making them sustainable and enduring, rather than focusing on sea–sun–sand and other common characteristics (Morgan *et al.*, 2002). Thus, harmonizing wine culture in place branding efforts can pave the way for sustainability in destinations with potential for wine tourism by activating cultural identity

*Corresponding author: nursahayhan@gmail.com

 Tourism Planning and Development in the Middle East (eds S. Kladou *et al.*)
DOI: 10.1079/9781789249149.0007

(Macionis, 1998). For this purpose, the place brand should comply with the core values of the place, including its inimitable wine culture and its stakeholders, so that mutual benefits and value are created for all (Buonincontri *et al.*, 2017; Jernsand and Kraff, 2017). Unfortunately, enfranchising economic aspects over societal prosperity in place branding causes failures when representing the place, its stakeholders and identity, and hinders sustainable tourism development (Mitropoulou and Spilanis, 2020; Rebelo *et al.*, 2020). This chapter analyses such aspects in the case of a specific wine island (Bozcaada) in Turkey. Recognizing the gap in describing the connection between place branding practice and sustainable development (Maheshwari *et al.*, 2011), this research aims to address and visualize their interconnectedness by laying its foundations on the identity-based approach to place branding. In response to its objective, this chapter highlights the importance of place identity, stakeholders' involvement and institutional support in the branding and tourism development process by exploring the perspectives of various local stakeholders on an island (Bozcaada/Tenedos) in Turkey. Turkey applies mostly top-down approaches in its tourism and branding strategies: being home to diverse geographies, minorities, culture and heritage attributes, the Turkish tourism offering presented in the Tourism Strategy – 2023 prioritizes some attributes over others (Ministry of Culture and Tourism, 2007), even if, as our study reveals, the chosen attributes are not accurately reflecting local (wine) identity. Interviews with local actors help us adopt a bottom-up approach to assess the island's culture and identity. Despite a strong potential for sustainable development building upon wine tourism, recent tourism development and branding 'efforts' disconnect the place culture from its past and sustainable potential.

The chapter is structured in five parts. First, the background is set using academic literature on sustainable tourism and wine culture in destination branding. The next section provides the readers with a description of Bozcaada and relevant recent regulations. The third section then explains the methods used for sampling and analysing and briefly describes the research sample. Findings in the fourth part highlight the views of local audiences of Bozcaada in relation to extant literature and the importance of wine and wine-related activities as integral to Bozcaada's identity and culture. Finally, the conclusion facilitates in-depth comprehension, and provides scholars and practitioners with ideas to inspire sustainability and dialogue for the benefit of stakeholders, culture and identity.

7.2 Literature Review

7.2.1 Sustainable tourism development and the identity-based approach to place branding

Sustainable tourism development enhances the living conditions of the local community, offers enriched experiences to tourists and sustains the environmental values (Yu *et al.*, 2009). Furthermore, sustainable tourism supports sustainable development by creating values for the local community (Liu, 2003). Unfortunately, destination marketing efforts often concentrate on consumers and image-building, and overlook internal audiences' values, which can enhance place identity (Wheeler *et al.*, 2011). Such practices ignore that sustainable brands emerge naturally, build on the meaning which locals attribute to their region and should then be integrated into consumer-driven marketing efforts (Wheeler *et al.*, 2011; Campelo *et al.*, 2014). Image and identity are two distinct but related constructs; image is the visible output of branding and marketing, whereas identity is essential for the content and success of them. Place image has two types: (i) projected image, as the promotion and presentation of the place to stimulate people; and (ii) perceived image, as the meaning and sense arising from people's internal activation process of these projected images (Bramwell and Rawding, 1996; Marine-Roig and Ferrer-Rosell, 2018). Institutions that have the 'power' to speak authoritatively are the ones that eventually conceptualize and 'shape' projected images (Deutsche, 1996; Smith, 2005), whereas perceived image takes form after exposure, according to each visitor's own characteristics. However, tourists and locals also influence perceived images since their opinions on social media can act as projected images and impact consumer decisions abruptly (Bernkopf and Nixon, 2019). Thus, depending on those practices, the image may be a narrowed, reified portrayal of the place rather than the multi-dimensionality of place identity.

Place identity can be (i) part of individual (human) identity; (ii) formative of group identity; (iii) individual's mental representations of the place; (iv) group perceptions of the place; (v) the identification of a group with a territory; and (vi) sense of place, 'character', 'personality' and distinctiveness (Kalandides, 2011, pp. 28–30). The static view of identity defines identities as internal elements that are disconnected from outsiders (Kavaratzis and Hatch, 2013); they are 'how we see ourselves' whereas images are 'how others see us'. 'Place identities are constructed through historical, political, religious and cultural discourses; through local knowledge, and influenced by power struggles' (Govers and Go, 2016, p. 17). As a result, place identity is subject to change over time, once the place and its local actors have embraced change. Unfortunately, authorities and tourism professionals may develop discursive place brands based on 'desirable' projected images, which locals then might seek to embrace thereby diluting their place identity (Broudehoux, 2017). In this context, sustainable development depends on the interconnection of locals' perception of their own land and the image bolstered by marketing efforts (Mitropoulou and Spilanis, 2020). Preserved culture and identity and honest representations of projected images escalate belongingness of residents as well as emotional attachment of visitors (Prayag and Ryan, 2012; Nocca, 2017), thereby the projected images could reflect the true sense of place to support sustainable development. At this point, place branding steps in to boost sustainable development by integrating local elements as the feasibility and sustainability of the place brand are not the outcome of closed boardroom meetings, but the inclusiveness of residents and culture in place branding efforts (Anholt, 2007; Gulisova, 2020). Considering the uptrend of bottom-up approaches to place branding (Kavaratzis, 2012; Mitropoulou and Spilanis, 2020), the identity-based approach can be a guideline to develop sustainable brands.

The identity-based approach highlights embedded culture as an overarching construct that both influences and is influenced by place identity (Kavaratzis and Hatch, 2013). Since place identity blends place culture and place branding (Kavaratzis and Hatch, 2013), culture, not solely composed of community members' individual attributes but also enhanced by the institutional framework it represents, is of great importance for any bottom-up work on developing sustainable place brands (Zouganeli *et al.*, 2012). As Kavaratzis and Ashworth (2015, p. 5) explain, culture works with place branding when we take culture as a provider of consumable (and saleable) experiences, a resource for economic activity, attracting the creative class and tourists and, lastly, an expression of the locality. In places where locality might relate to grape harvesting, viticulture and other similar aspects, the importance of wine culture for place branding emerges.

7.2.2 Building upon wine culture for sustainable tourism development and place branding

Sustainable development builds upon four pillars: economic, social, environmental and cultural (World Commission on Environment and Development, 1987; Nurse, 2006; United Nations, 2012, 2016; Nocca, 2017). Nocca (2017) highlights that gastronomy stands for social and economic improvement, embodies local culture, emphasizes identity and evokes visitors. Moreover, gastronomy is a cultural concept with figurative and authentic elements, rather than a practical offering (Neal, 2006). Wine, a component of 'gastronomic identity', is an outcome of cultural interactions and traditions (Harrington, 2005), whereas wine tourism deeply accompanies uniqueness of local culture and identity (Montella, 2017). Thus, sustainable wine development should be achieved through sustainability pillars: engaging local cultural elements, expanding visitor experiences, enhancing collaboration among stakeholders and preserving the resources for wine production (Poitras and Donald, 2006).

Place branding expresses destinations' uniqueness to outsiders while strengthening the local identity within the internal forces (Colomb and Kalandides, 2010). Wine is a definitive product of places with a long wine culture and tradition, constructed with the engagement of consumers and locals (Zaichkowsky, 1985; Thode and Maskulka, 1998). A well-developed and protected wine industry can catalyse regional and cultural identity and boost a specific type of tourism (Macionis, 1998). Harmonizing wine tourism with the overall destination experience will then bring about an enlarged projected image (Hall *et al.*, 2002) and align the image with identity.

McKercher (2002, pp. 32–33) classified cultural tourists as (i) 'purposeful cultural tourists' who intend to deeply experience the culture, (ii) 'sightseeing cultural tourists' who want to learn about the culture at first but end up with shallow experiences, (iii) 'casual cultural tourists' who attach medium importance to culture and have low - intensity experiences, (iv) 'incidental cultural tourists' who do not have cultural motivations and leave with shallow experiences, and (v) 'serendipitous cultural tourists' who do not have cultural concerns but leave the destination with intense experiences. 'Wine tourism', a cultural tourism type, is driven by the visitors' motivations of experiencing wine regions, visiting vineyards, wineries and wine-related events (Hall *et al.*, 2002). Johnson (1998, p. 15) identified wine tourists as 'specialist winery tourists', who are specifically interested in grape wines and visiting wineries, wine-related festivals, and 'generalist wine tourists', who have other motivations to visit. Wine-related festivals are significant, especially because they affect the attractiveness of the regions in the eyes of wine tourists (Houghton, 2001). On the other hand, the recent projected images of wine destinations are more experience-based and focus primarily on surroundings, nature, culture and other local characteristics to build up remarkable memories for visitors, thereby better matching with the purposeful cultural tourist profile and emphasizing the importance of connecting local identity with projected images to favour perceived image traits (Williams, 2001a, b; Alant and Bruwer, 2004). When wine is embedded in the local identity, developing wine tourism experiences may inspire co-creation among wine-related stakeholders and visitors (of wine festivals, wineries, wine restaurants, vineyards, the wine destination etc.).

The literature usually connects the branding process with customers and image studies, while co-creation expands to include other stakeholders too (Hatch and Schultz, 2010). 'Enterprise branding' is the idea that a brand represents both the organization and all stakeholders involved in its intentions and activities (Hatch and Schultz, 2009, 2010). Similarly, 'inclusive place branding' embraces the concept of involvement and enablement of stakeholders, as the traditional methods underestimate the multiplicity of place and culture for long-term sustainability (Kavaratzis *et al.*, 2017). Service-dominant logic of marketing places local actors (such as tourism and place authorities, organizations, associations, businesses, travellers and locals), their dialogues and institutional networks at the centre of any place branding exercise in order to benefit involvement and representation of the 'sense of place' and place identity, while co-creating diversified and cultural value for all (Brown, 2017; Giannopoulos *et al.*, 2020).

Thus, integrating local actors (local, provincial, national governments, residents, viniculture, viticulture and wine professionals, local business owners and tourism administrators) and fundamental support of authorities are crucial for sustainable wine destinations, inspire sustainability and improve culture-based tourism (Skinner, 2002; Tomljenović, 2009). Due to the inevitable impacts of branding actions on inhabitants, residents should be included as key stakeholders during co-creation (Kavaratzis and Hatch, 2013). However, top-down tourism planning efforts overlook local co-creators and their identity or consider certain representatives that prioritize some stakeholders' interests (Eshuis *et al.*, 2013; Zenker and Erfgen, 2014; Jernsand and Kraff, 2017). This approach may result in alienation and limited incorporation of the real essence within the place brand (Kavaratzis, 2012), although place branding should transcend economic benefits and represent also under-privileged stakeholders for sustainable development (Kavaratzis *et al.*, 2017; Rebelo *et al.*, 2020). The developments in Bozcaada, Turkey, hereby emerge as an interesting case illustrating how branding efforts contradict the identity-based approach to place branding and overshadow sustainable development.

Fig. 7.1. Location of Bozcaada in Turkey. Adapted with permission from MapSVG, 2021.

7.2.3 The case of Bozcaada in Turkey

With a long coastline, and countless cultural and natural attractions (Cooper and Ozdil, 1992), Turkey is a Eurasian country with nearly 84 million people (Worldometers, 2020). The Tourism Encouragement Law of 1982 pioneered tourism investments on the shores of the Mediterranean and the Aegean Sea and boosted a monotype (sun–sea–sand) mass tourism development, thereby largely ignoring distinctive regional/local characteristics (Göymen, 2000; Tosun, 2001; Tosun *et al.*, 2003; Kuvan, 2010). More recently, the Tourism Strategy of Turkey – 2023 (Ministry of Culture and Tourism, 2007) and the 11th Development Plan (Republic of Turkey, 2019) set objectives more relevant to sustainable development and priorities towards alternative tourism types (e.g. winter tourism, cultural tourism and gastronomy tourism).

Bozcaada (Tenedos) is a small island located in the Çanakkale province in Turkey (see Fig. 7.1). Although, the population can reach up to 10,000 in the summer (inclusive of people with summer homes and visitors), the permanent population was 3023 people in 2018 (Official Website of Bozcaada, n.d). Until the second half of the 20th century, the population was mostly composed of *Rums* (i.e. a Greek speaking Orthodox community, holding a Turkish or dual Turkish–Greek nationality), yet the more detached the *Rums* felt, the more they started to leave Bozcaada (Gürüney, 2009). Recently, the social structure has changed; the population born in Çanakkale declined from 69% to 58%, whereas the ones from Istanbul increased to 11% from 3%, between 1990 and 2014 (TUIK, 2015). When the majority was *Rums*, the local economy was dependent on viticulture and wine production activities with vineyards covering one-third of the island and 13 wineries in total; however, nowadays, it is tourism-driven and six wineries remain: Corvus, Gülerada, Amadeus, Yunatçılar, Ataol and Talay (Akpinar *et al.*, 2011; Çavuşoğlu, 2012; Okumuş, 2018). The production capacity nowadays reaches up to 6000 tonnes, accounting for 5–10% of Turkey's wine production (Türker and Alaeddinoğlu, 2016). Both endemic wine grapes (Karasakız (Kuntra), Altınbaş (Vasilaki) and Karalahna) and well-known French grapes (Cabernet Sauvignon, Merlot, Gamei and Shordone) grow on the island (Akpinar *et al.*, 2011, p. 339). In terms of wine tasting, Yunatçılar and Talay have founded wine bars in the 'city centre' for this purpose, Amadeus and Gülerada offer wine tasting at their winery premises, and Corvus both in its sales office and winery. Talay, Corvus and Amadeus also offer vineyard and winery visits. Bozcaada's wine and vineyard tourism could diversify the island as a sustainable destination and attract more 'special interest tourists' as it would be unjust for Bozcaada to be noted only for sun–sea–sand (Çavuşoğlu, 2012).

In terms of the cultural events of Bozcaada, the role of wine is still central. The annual Grape Harvest Festival, Local Tastes Festival, Bozcaada International Salhane Wine Days and International Wine Competition promote wine culture and locality. The Day of Aya Paraskevi and Ayazma Fair (i.e. a traditional food and wine festival) are great examples of the *Rum* heritage and mixed ethnic (and religious) culture of Bozcaada. Besides wine- and agriculture-related activities, Bozcaada also encourages alternative festivals (Bozcaada Jazz Festival, Bozcaada International Festival of Ecological Documentary and Bozcaada Marathon).

Sustainable tourism development is 'circular' as it incorporates 'human and institutional change' and creates prosperity for all pillars of sustainability – environmental, economic and socio-cultural – in the short and long term, as well as improving residents' quality of life (Hapgood, 1969, p. 20; Tosun, 2001; Nocca, 2017). On the other hand, mass tourism eventually endangers ecosystems, indigenous culture and nature, and is at odds with wine tourism (Poitras and Donald, 2006; Kuvan, 2010). In Turkey, however, local authorities are under the control of central authorities and local communities are mostly excluded from the decisions; thus, planning often disregards the interests and culture of local people (Ersoy, 1992; Tosun, 2001).

The Tourism Strategy of Turkey – 2023 discusses the collaboration of central and local governments, civil actors' involvement in decision making and supports alternative tourism types to diffuse tourism (Ministry of Culture and Tourism, 2007). The development plan (Republic of Turkey, 2019) also provides some support to vineries. Even though alcohol existed in Anatolia as a part of culture before Turks and Islam, recently it has been debated based on the priorities of each ruling party. The Tobacco and Alcohol Market Regulatory Authority (TAPDK), established in 2002, led to increased control over society's consumption of alcohol. In 2013, the Global Alcohol Policy Symposium criticized 'rakı' for being promoted as Turkey's national drink, and suggested 'ayran' (i.e. a cold yogurt-based beverage) instead (Evered and Evered, 2016). Since 2013, new regulations constrain the advertisement of alcoholic beverages and prohibit any campaign, promotion or event that encourages and incentivizes the consumption and sale of such products (such as wine festivals or open wine tastings) as well as sponsorship of such events. Given the challenges for wine culture and tourism posed by such regulations, this study focuses on Bozcaada to explore the dialogical relationship between sustainable tourism development and branding efforts.

7.3 Methodology

This study is the first to explore the interlink between wine tourism and wine culture in Bozcaada. Given this exploratory character, a qualitative approach helps us better explore the place brand, destination identity and the wine culture of Bozcaada through the eyes of its local stakeholders. Specifically, we held semi-structured interviews to investigate stakeholders' perceptions and their willingness to co-create a sustainable place brand and examine the synergies between the identity-based approach and sustainable tourism development. Grounded theory led the data collection and analysis (Goulding, 1999).

As place brands are formed through the contribution of authorities and local stakeholders (Giannopoulos *et al.*, 2020), this study draws from the views of residents, wine producers, local business owners and the local municipal authority. Data collection was undertaken on Bozcaada in April 2019. In qualitative sampling, the sample's ability to represent important attributes is significant (Ritchie *et al.*, 2014). In this regard, the experts' sample was purposively selected according to their relationship to culture and/or tourism on the island, occupation and the years they have spent on Bozcaada. Purposive sampling relies on researchers' *a priori* theoretical understanding and the assumption of particular individuals' adequate knowledge and important views on the phenomenon under study (Robinson, 2014). We first contacted the local public authority, all winery owners and relevant business owners, but not all winery and business owner replied to the call. Then we randomly approached residents in local hospitality establishments, eventually leading to the sample profiled in Table 7.1. The invitation informed participants of the purpose of the study and the way in which data would be used. In total, we conducted 16 semi-structured interviews with 4 wine producers, 6 local business owners, 5 residents and the single local authority. Each interview lasted between 45 min and 1 h and

Table 7.1. Sample of the research.

Stakeholder Group[a]	Age Group	Years of residence in Bozcaada	Occupation
Wine producer	61+	30	Wine producer
Wine producer	61+	Born & raised	Wine producer
Wine producer	51–60	Born & raised	Wine producer
Wine producer	51–60	25	Wine producer
Local business Owner	41–50	15	Hotel owner
Local business Owner	41–50	Born & raised	Restaurant owner
Local business Owner	31–40	3	Coffee shop owner
Local business owner	41–50	9	Hotel owner
Local business owner	41–50	12	Ceramics shop owner
Local business owner	31–40	4	Coffee shop owner
Resident	61+	59	Jeweller, former winery worker
Resident	41–50	20	Electrician
Resident	61+	57	Retired winery worker
Resident	61+	Born & raised	Housewife
Resident	61+	Born & raised	Retired winery worker
Local authority	41–50	16	Mayor

[a] The term 'wine producer' refers to the professionals managing one of the local wineries, whereas the term 'local business owner' accredits the owners of the bar, cafe, restaurants and accommodation facilities.

included 27 questions, proceeding from 'grand tour' questions (McCracken, 1988) to explaining (i) the current situation for wine tourism, (ii) previous and current purposes of visitors, (iii) the (wine) culture of the island, (iv) appropriate strategies for the brand of Bozcaada regarding wine culture, and (v) the impact (if any) of laws and regulations. We observed that the data were adequate with participants referring to the same themes as in previous interviews, thereby data collection for residents too stopped when we reached the saturation point (Alsaawi, 2014).

As Table 7.1 illustrates, wine producers are over 51 years old, two of them were born and raised on the island, but the other two have also spent nearly half of their lives on Bozcaada. One wine producer was an architect before getting involved in wine making and another one was a chemical engineer. The wine producers who were born on the island continue the longstanding wine tradition of their families, with one being from the first Muslim family to produce wine on Bozcaada. The local business owners are younger, aged between 31 and 50. Only one local business owner was born and raised on the island, whereas the others have resided on Bozcaada for 3, 4, 9, 12 and 15 years. Two of them own hotels, two others own coffee shops, one has a restaurant and one has a ceramics shop. The mayor of the island is a doctor, whereas the residents have a diversified profile, one being an old *Rum* housewife, another an electrician, and two previously worked in wineries but now one of them is a jeweller and the other is retired. The electrician has the least 'experience' residing on Bozcaada: 20 years – almost his entire adult life, that is.

After data collection, we checked the transcripts for accuracy, then used thematic analysis (Braun and Clarke, 2006). Following several rounds of open coding, we reached theoretical understanding and identified main emerging topics, then we restructured interrelated themes and subthemes by re-organizing and comparing thematic categories to refine the data under each theme. The main emerging themes related to wine (including agriculture, wine culture and wine tourism), the identity-based approach (embracing branding, Greek culture and co-creation) and sustainable tourism development (referring to visiting motives, socio-cultural aspects relevant to the local identity and economic aspects relevant to sun–sea–sand versus wine tourism). Our findings, presented in the following section, illustrate the interconnections among these recognized themes.

7.4 Findings

7.4.1 Branding Bozcaada, the identity-based approach and the role of wine

At first, interviewees discussed their perceptions of the identity and culture of Bozcaada and provided us with the first points to build the framework describing the synergies between sustainable tourism development and the identity-based approach to place branding (Fig. 7.2). Although we asked about culture in general, respondents immediately referred to agriculture, wine, viticulture and viniculture. Gastronomic identity and wine represent cultural communications and authenticity (Harrington, 2005; Nocca, 2017). Interestingly, Greek/*Rum* culture – or nostalgia about it – was still dominant among all interviewees despite the decrease in the *Rum* population and the cultural change. The harmony of two different ethnic groups and religions was a distinctive aspect of Bozcaada's identity (Akpinar *et al.*, 2011). Culture is a process linking the place brand with locals when it is preserved (Kavaratzis and Ashworth, 2015; Nocca, 2017); however, in this case, there was a disconnection between the culture and locals in Bozcaada. One respondent stated: 'while the main culture should reflect the Greek culture, those who came here after Greeks left were abusers of this culture' (male, 62, wine producer). As wine indicates cultural authenticity (Thode and Maskulka, 1998), respondents were of one mind that Bozcaada is an agricultural, mainly 'wine island' with nearly 4000 years of wine culture: 'wine was and still is one of the most important elements of (local) culture the island, Bozcaada should never be a (mass) tourism island, it is an agricultural island, and this agricultural island hosts a wine industry' (male, 62, wine producer); 'almost everyone owns or is related to somebody who owns a vineyard, everyone is dealing with the vineyard in a way that links to wine' (male, 47, local business owner).

Natural beauty emerged as important for stakeholders, and several interviewees used *'isolation'* and *'being an island'* while referring to Bozcaada's culture and identity. Islands are well-known destinations for attracting tourists in search of seclusion and natural beauty (Hall, 2010). Informants mentioned nature, weather, sea and tranquillity as distinctive characteristics of Bozcaada. Within this context, in the past, tourists were motivated by natural characteristics and (wine) culture. Surrounding environment, nature, ambiance and culture are significant to create an attractive wine tourism destination as the visual qualities of the local environment are the determinatives of culture (Alant and Bruwer, 2004; Kavaratzis and Ashworth, 2015). In the past, tourists were coming for wine tasting and wine experiences, whereas currently, popularity is their primary motivation: 'In the past, people who were familiar with Bozcaada preferred the island for its nature, its cultural fabric, but now people come because it is popular, trendy' (male, 45, local business owner). In the case of Bozcaada, identity, culture and image have scattered apart from each other, thus have changed visitors' motivation and residents' approach to their place, along with the socio-cultural change due to increased popularity and migration from cities, especially Istanbul (Okumuş, 2018).

7.4.2 Visiting motives, tourism profiles and their link with (un)sustainable tourism development

All stakeholders mentioned visitors came to experience the island's wine culture, taste the wines and participate in wine-related activities in the past, as purposeful cultural tourists, and described current visitors as excursionists, or, at best, incidental or casual cultural tourists, as they mostly visit the island for sun–sea–sand and because of its popularity (McKercher, 2002). For example, an informant stated that '10 years ago [i.e. around 2010] the intellectual tourists stopped coming to the island; that is such a loss. Now standard, monotype tourists are those who visit' (male, 47, local business owner). In this case, when the *Rums* left the

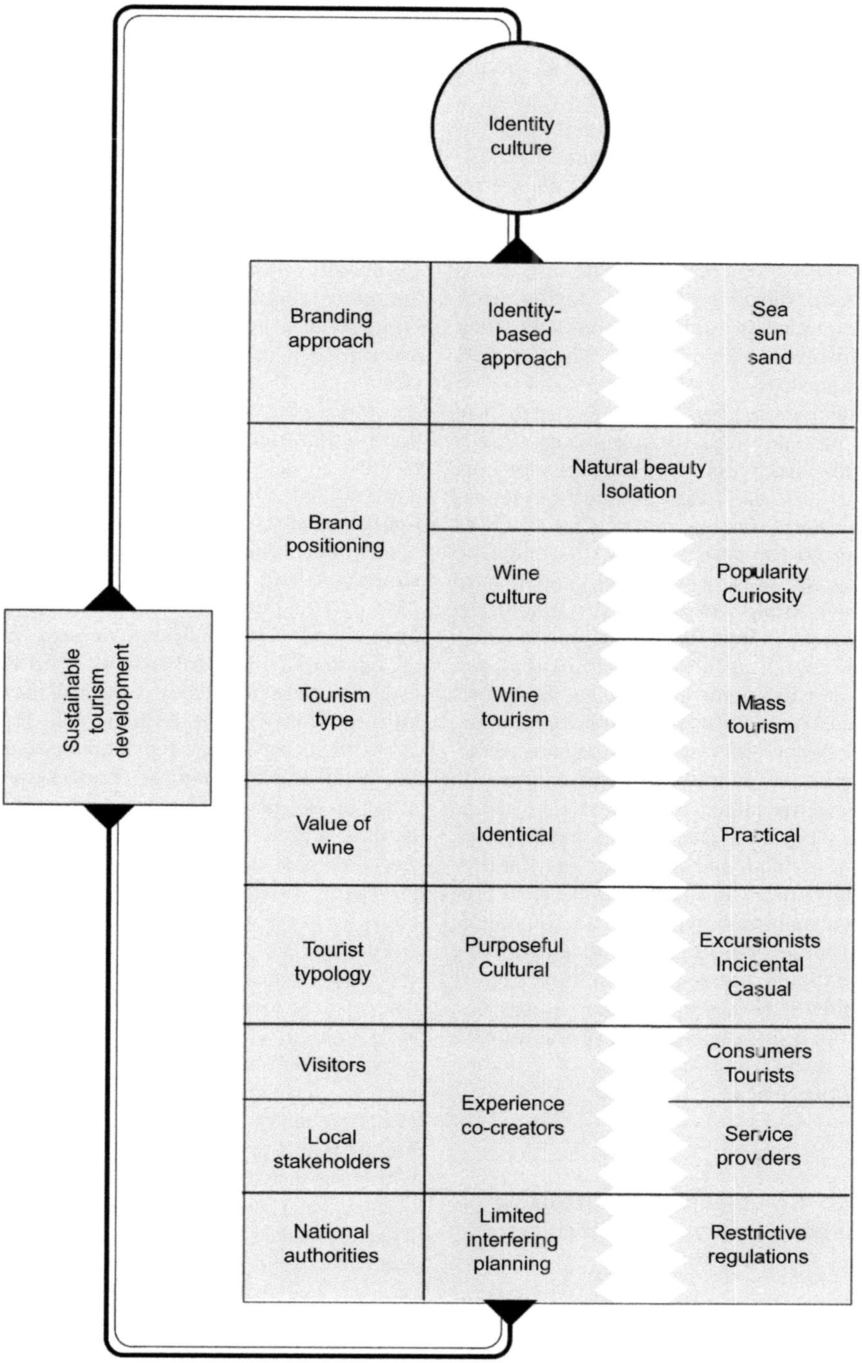

Fig. 7.2. Developing synergies between the identity-based approach to place branding and sustainable tourism development in the case of wine tourism.

island, Turks replaced the services according to what tourists might like, instead of reflecting the real culture. Unfortunately, abrupt identity transition and popularity affected the tourist profile, tourism activities and local identity, caused a socio-cultural alienation on the island and led to 'a region without identity' (Akpinar *et al.*, 2011, p. 334). One informant concluded 'We are no longer reflecting our identity to tourists; instead, we are keeping up with their identity' (male, 35, local business owner). However, sustainable brands take shape organically, starting with the meaning locals ascribe to their place, then getting involved in place image (Wheeler *et al.*, 2011; Campelo *et al.*, 2014).

Evidently, wine has a crucial place in local people's everyday life. All stakeholders agreed that locals largely consume alcohol, not only wine but also 'rakı'. One of the interviewees indicated '*Rums* were so hardworking, but they never missed the parties in town, so now we are doing the same way by working so hard but always drink wine and have fun in the afternoon, that's how island lifestyle should be' (male, 47, local business owner). In the past, tourists consumed more wine, but nowadays they at least buy wine as souvenir before leaving, even if they have not tasted it, visited any wineries or expressed any interest in wine culture. As one interviewee stated: 'The tourists buy wine as a gift. They hear that grapes and wine are parts of the island, and view wine as what they should buy from the island, not because they know that we have a [wine] culture' (male, 35, local business owner). However, gastronomy and wine should represent local culture, identity and authenticity, thereby attracting tourists, rather than being a non-subjective and tangible outcome during and at the end of visitation (Neal, 2006; Nocca, 2017).

7.4.3 Co-creation for sustainable tourism development: a focus on institutions

With respect to the role of institutions and the intention to co-create the destination brand, both similarities and differences were detected in the opinions of wine producers, local business owners and residents. One wine producer was eager to inspire co-creation by combining agriculture, wine and tourism, whereas the rest preferred focusing solely on wine production. Their limited intentions resulted from ignorance, limited knowledge on how to co-create, as well as their belief that other local businessmen seek only to maximize their own profit, and co-creation should first be pursued by state authorities. Some local business owners also said the ignorance and monetary expectations of some stakeholders are challenges; for example, 'there are many ignorant stakeholders, [...] because of seasonality many of the business owners overrun their capacity with lower quality services' (male, 35, local business owner).

Since wine tourism and wine-related activities are thoroughly related to local culture and identity, to develop sustainable wine tourism destinations, harmonizing local culture, upgrading visitor experience, promoting stakeholder interaction, and conserving resources for viniculture and wine production are essential (Poitras and Donald, 2006; Montella, 2017). Most interviewees indicated negative impacts of regulation changes on the activities and festivals representing the culture of Bozcaada and shared negative feelings and thoughts about them, and suggested such events should be revitalized: 'Regulation changes were too much for Bozcaada; free wine-tasting days were banned, and that influenced everything. I wish we could take all back' (male, 64, resident). Restricting wine-related festivals has posed obstacles to embracing and sustaining the wine culture of Bozcaada, since festivals increase the attractiveness of wine regions, encourage tourists to be involved and preserve culture as it is (Houghton, 2001). In response, interviewees suggested wineries should co-organize tourists' participation in harvesting grapes, winery visitation and wine tasting. As a respondent highlighted: 'this way you can still do something within the legal framework. The vintage festival is officially forbidden. I don't know why because it doesn't make sense; there is no problem when you collect grapes for your table, but there is a problem when you make wine?!' (male, 55, wine producer).

Limited governmental support and the lack of a common vision appeared as significant reasons preventing co-creation. Wine producers and the local authority were distinguished from local business owners and residents by directly identifying laws, regulations and current poor overall service quality as barriers to branding efforts. Almost all stakeholders expressed frustration over the barriers to development and new laws and regulations. Almost all were disappointed at not having the opportunity to engage in brand-development decisions, and not seeing decision makers' intentions to involve them. To overcome these barriers, local business owners and residents proposed an interactive cooperation among local stakeholders of Bozcaada and invited state authorities to initiate and encourage such cooperation. Whereas the local authority advocated a tourism masterplan to co-create Bozcaada, most local business owners and residents suggested that a sustainable wine destination brand can be attached to Bozcaada if the government supports locals in terms of wine production, promotion and wine-related events. Place branding is co-managed by a multitude of place-makers (Kavaratzis, 2012), yet, in Bozcaada, most of the stakeholders do not believe they can work together if central authorities do not encourage them. Thus, central authorities should support and initiate progress of sustainable tourism development for the well-being of the destination, its culture and identity (Göymen, 2000; Tosun, 2001). Besides, inclusive place branding stands for the multiangular nature of place, identity and culture, and prioritizes collective efforts by internal stakeholders and policy makers in a way that ensures all stakeholders are represented to bolster cultural tourism and sustainable development (Tomljenović, 2009; Kavaratzis *et al.*, 2017; Rebelo *et al.*, 2020).

7.5 Conclusions

Given the identity-based approach to place branding (Kavaratzis and Hatch, 2013), this study has explored local stakeholders' perspective on Bozcaada's identity, (wine) culture and the place brand. Academic communities may benefit from the case by analysing the provided framework depicting the dialogical relationship between sustainable tourism development and the identity-based approach to place branding, and how they nourish each other in the case of wine tourism. Experience plays a central role in redeveloping sustainable brands, and for memorable experiences, interest from the co-creation parties is essential (Bruwer and Alant, 2009). To develop engaging destination experiences and valuable place brands, the image and culture of the place should be matched (Buonincontri *et al.*, 2017). Viticulture and wine culture that have been fundamentally constructed by *Rum* residents and lately imparted to Turks, along with nature and isolation, were underlined when defining the identity of Bozcaada. Wine still holds a significant place in locals' lifestyles; high levels of consumption and production continue, whereas tourists' interest and involvement has decreased. Thus, the importance of the practical implications outlined below emerge.

Managers and policy makers could use our framework to understand their own position towards sustainable development and look at how institutions may influence co-creation. The migration of people who escaped from 'big city life' (Okumuş, 2018, p. 117) and 'pressures of the modern world' (Smith and Phillips, 2001, p. 464) has influenced the socio-cultural environment and place brand of Bozcaada. Promotion and consumption of wine and wine-related events are stymied by laws and regulations; therefore, they have become hidden values and expressions of culture on Bozcaada. We argue that the co-creation process can still be achieved. Although wine producers seemed reluctant, other interviewees were more willing if institutional support is provided. In Turkey, cultural tourism is encouraged and complements sun–sea–sand tourism (Okumus *et al.*, 2012), yet wine tourism is not seen as a form of cultural tourism. National and regional plans seek to prioritize alternative tourism development, encourage agriculture and viniculture activities and include gastronomy tourism highlighting Bozcaada's endemic Çavuş grape (see the 11th Development Plan, Republic of Turkey, 2019 and the Çanakkale Tourism Investment Guide, South Marmara Development Agency, 2018). Still, wine tourism is not mentioned as an element of gastronomy tourism either. The plans are in accordance with the regulations on alcoholic beverages even if such tourism planning and development efforts contravene sustainable tourism development, local place identity and culture.

This chapter started by arguing that sustainable development and place branding need to be interlinked in order for both to be effective in practice. Such bundled approaches are of distinct importance in the case of Bozcaada and the COVID-19 aftermath. Our analysis revealed the significance of isolation, nature and wine for the brand identity of the island. Social distance (aka isolation) and experiences in nature gain more meaning because of the pandemic; among others, Spalding *et al.* (2021) highlight the contribution of nature, its economic and non-economic value and the role of nature in the post-pandemic tourism recovery. What is more, post-lockdown tourism intentions favour activities linked to gastronomic and agricultural activities, including viticulture and wine (Ateljevic, 2020; Gastaldello *et al.*, 2022). However, sustainable tourism is plausible when resilience is built on government response and local belongingness (Sharma *et al.*, 2021), thereby establishing the link between identity-based place branding and sustainable tourism policy and development.

The central structure of decision making leads to the central government being the one controlling tourism development, thus causing incompatibility between privileged stakeholders' benefits and locals' values (Seckelmann, 2002). On the other hand, sustainable tourism development can be achieved by engaging identity and culture, local involvement and cooperation. Protecting the distinctive characteristics of Bozcaada rather than simply degenerating the island to a regular sun–sea–sand destination and engaging various stakeholders in branding could result in worthwhile and reinforced place brand value, so the island can be revived as a culturally coherent destination. Locals transforming their houses into accommodation complexes and letting tourists invade their private lives can degrade authenticity, degenerate and even assimilate the local culture (Okumuş, 2018; Zhou *et al.*, 2018). Major improvements can be achieved if branding efforts are focused on collective and sustainable growth instead. The government should support and represent local stakeholders as well as the culture and identity to provide the island a sustainable brand opportunity. Towards this direction, traditional methods that the *Rum* population used to incorporate can be bundled together with a joint effort to brand the island as a wine destination. Similarly, brand experts need to consider prioritizing local rather than national culture and setting a framework within which the dialogical relationship between place branding and sustainable tourism development will flourish.

This exploratory study is a first step towards examining wine tourism on Bozcaada. Future research needs to go into more detail and include the perspective of more residents, central and national authorities together with that of tourists. Furthermore, there is a systematic harassment of the cultural texture and even daily life in Bozcaada, resulting from the restrictive practices of authorities. The laws and regulations about alcoholic beverage production, promotion and consumption should be investigated as a new research subject analysing how institutions change, maintain and decay, thereby impacting development, sustainable place branding and identity. Still, we believe that, by generating significant implications from the stakeholders involved and data collected, this research offers new knowledge to the scope of research and promotes further studies investigating the interconnection between sustainable tourism development and place branding.

References

Akpinar, F., Saygin, N. and Karakaya, E. (2011) Evaluation of the conservation activities in the historical settlement Tenedos-Bozcaada Island. In: Brebbia, C.A. and Beriatos, E. (eds) *Sustainable Development and Planning V*. WIT Press, Southampton, UK, pp. 329–343. DOI: 10.2495/SDP110281.

Alant, K. and Bruwer, J. (2004) Wine tourism behaviour in the context of a motivational framework for wine regions and cellar doors. *Journal of Wine Research* 15(1), 27–37. DOI: 10.1080/0957126042000300308.

Alsaawi, A. (2014) A critical review of qualitative interviews. *European Journal of Business and Social Sciences* 3(4), 149–156. DOI: 10.2139/ssrn.2819536.

Anholt, S. (2007) Nation-brands and the value of provenance. In: Morgan, N., Pritchard, A. and Pride, R. (eds) *Destination Branding: Creating The Unique Destination Proposition*. Elsevier, Oxford, UK, pp. 26–39.

Ateljevic, I. (2020) Transforming the (tourism) world for good and (re)generating the potential 'new normal'.' *Tourism Geographies* 22(3), 467–475. DOI: 10.1080/14616688.2020.1759134.

Bernkopf, D. and Nixon, L. (2019) The impact of visual social media on the projected image of a destination: the case of Mexico City on instagram. In: Pesonen, J. and Neidhardt, J. (eds) *Information and Communication Technologies in Tourism*. Springer, Cham, Switzerland, pp. 145–157. DOI: 10.1007/978-3-030-05940-8_12.

Blain, C., Levy, S.E. and Ritchie, J.R.B. (2005) Destination branding: insights and practices from destination management organizations. *Journal of Travel Research* 43(4), 328–338. DOI: 10.1177/0047287505274646.

Bramwell, B. and Rawding, L. (1996) Tourism marketing images of industrial cities. *Annals of Tourism Research* 23(1), 201–221. DOI: 10.1016/0160-7383(95)00061-5.

Braun, V. and Clarke, V. (2006) Using thematic analysis in psychology. *Qualitative Research in Psychology* 3(2), 77–101. DOI: 10.1191/1478088706qp063oa.

Broudehoux, A.M. (2017) *Mega-Events and Urban Image Construction: Beijing and Rio de Janeiro*, 1st edn. Routledge, London. DOI: 10.4324/9781315393308.

Brown, J. (2017) A service ecosystem approach to representing a place's unique brand. In: Kavaratzis, M., Giovanardi, M. and Lichrou, M. (eds) *Inclusive Place Branding: Critical Perspectives on Theory and Practice*. Routledge, Oxford, UK. DOI: 10.4324/9781315620350-5.

Bruwer, J. and Alant, K. (2009) The hedonic nature of wine tourism consumption: an experiential view. *International Journal of Wine Business Research* 21(3), 235–257. DOI: 10.1108/17511060910985962.

Buonincontri, P., Morvillo, A., Okumus, F. and van Niekerk, M. (2017) Managing the experience co-creation process in tourism destinations: empirical findings from Naples. *Tourism Management* 62, 264–277. DOI: 10.1016/j.tourman.2017.04.014.

Campelo, A., Aitken, R., Thyne, M. and Gnoth, J. (2014) Sense of place: the importance for destination branding. *Journal of Travel Research* 53(2), 154–166. DOI: 10.1177/0047287513496474.

Çavuşoğlu, M. (2012) Bozcaada grape farming tourism and electronic tourism design site application. *International Journal of Social and Economic Sciences* 2(2), 49–54.

Colomb, C. and Kalandides, A. (2010) The 'be Berlin' campaign: old wine in new bottles or innovative form of participatory place branding. In: Ashworth, G. and Kavaratzis, M. (eds) *Towards Effective Place Brand Management: Branding European Cities and Regions*. Edward Elgar Publishing, Cheltenham, UK and Northampton, Massachusetts, pp. 173–190. DOI: 10.4337/9781849806398.00016.

Cooper, C.P. and Ozdil, I. (1992) From mass to 'responsible' tourism: the Turkish experience. *Tourism Management* 13(4), 377–386. DOI: 10.1016/0261-5177(92)90005-R.

Deutsche, R. (1996) *Evictions: Art and Spatial Politics*. MIT Press, Cambridge, Massachusetts.

Ersoy, M. (1992) Relations between central and local governments in Turkey: an historical perspective. *Public Administration and Development* 12(4), 325–341. DOI: 10.1002/pad.4230120402.

Eshuis, J., Braun, E. and Klijn, E.H. (2013) Place marketing as governance strategy: an assessment of obstacles in place marketing and their effects on attracting target groups. *Public Administration Review* 73(3), 507–516. DOI: 10.1111/puar.12044.

Evered, E.Ö. and Evered, K.T. (2016) From *rakı* to *ayran*: regulating the place and practice of drinking in Turkey. *Space and Polity* 20(1), 39–58. DOI: 10.1080/13562576.2015.1057003.

Gastaldello, G., Livat, F. and Rossetto, L. (2022) Does covid scare wine travelers? Evidence from France and Italy. *Wine Economics and Policy*. Available at: https://oaj.fupress.net/index.php/wep/article/view/11550 (accessed 25 October 2022).

Giannopoulos, A., Piha, L. and Skourtis, G. (2020) Destination branding and co-creation: a service ecosystem perspective. *Journal of Product & Brand Management* 30(1), 148–166. DOI: 10.1108/JPBM-08-2019-2504.

Giovanardi, M., Lucarelli, A. and Pasquinelli, C. (2013) Towards brand ecology: an analytical semiotic framework for interpreting the emergence of place brands. *Marketing Theory* 13(3), 365–383. DOI: 10.1177/1470593113489704.

Goulding, C. (1999) Grounded theory: some reflections on paradigm, procedures and misconceptions. Working Paper Series, Management Research Centre. University of Wolverhampton, Wolverhampton, UK.

Govers, R. and Go, F. (2016) *Place Branding: Glocal, Virtual and Physical Identities, Constructed, Imagined and Experienced*. Palgrave Macmillan, Basingstoke, UK.
Göymen, K. (2000) Tourism and governance in Turkey. *Annals of Tourism Research* 27(4), 1025–1048. DOI: 10.1016/S0160-7383(99)00127-9.
Gulisova, B. (2020) Rural place branding processes: a meta-synthesis. *Place Branding and Public Diplomacy* 17(4), 368–381. DOI: 10.1057/s41254-020-00187-y.
Gürüney, M.H. (2009) *From Tenedos to Bozcaada: Tale of Forgotten Island*. Bozcaada Local History Research Center, Çanakkale, Turkey.
Hall, C.M. (2010) Island destinations: a natural laboratory for tourism: introduction. *Asia Pacific Journal of Tourism Research* 15(3), 245–249. DOI: 10.1080/10941665.2010.503613.
Hall, C.M., Longo, A.M., Mitchell, R. and Johnson, G. (2002) Wine tourism in New Zealand. In: Hall, C.M., Sharples, L., Cambourne, B. and Macionis, N. (eds) *Wine Tourism Around the World: Development, Management and Markets*. Elsevier, Oxford, UK, pp. 150–174.
Hapgood, D. (1969) *The Role of Popular Participation in Development*. MIT Press, Cambridge, Massachusetts.
Harrington, R.J. (2005) Defining gastronomic identity. *Journal of Culinary Science & Technology* 4(2–3), 129–152. DOI: 10.1300/J385v04n02_10.
Hatch, M.J. and Schultz, M. (2009) Of bricks and brands: from corporate to enterprise branding. *Organizational Dynamics* 38(2), 117–130. DOI: 10.1016/j.orgdyn.2009.02.008.
Hatch, M.J. and Schultz, M. (2010) Toward a theory of brand co-creation with implications for brand governance. *Journal of Brand Management* 17(8), 590–604. DOI: 10.1057/bm.2010.14.
Houghton, M. (2001) The propensity of wine festivals to encourage subsequent winery visitation. *International Journal of Wine Marketing* 13(3), 32–41. DOI: 10.1108/eb008725.
Hultman, J. and Hall, C.M. (2012) Tourism place-making: governance of locality in sweden. *Annals of Tourism Research* 39(2), 547–570. DOI: 10.1016/j.annals.2011.07.001.
Jernsand, E.M. and Kraff, H. (2017) Democracy in participatory place branding: a critical approach. In: Kavaratzis, M., Giovanardi, M. and Lichrou, M. (eds) *Inclusive Place Branding: Critical Perspectives on Theory and Practice*. Routledge, London, pp. 53–65. DOI: 10.4324/9781315620350-2.
Johnson, G.R. (1998) Wine tourism in New Zealand: a national survey of wineries. PhD thesis, University of Otago, Dunedin, New Zealand.
Kalandides, A. (2011) The problem with spatial identity: revisiting the 'sense of place'. *Journal of Place Management and Development* 4(1), 28–39. DOI: 10.1108/17538331111117142.
Kavaratzis, M. (2012) From'necessary evil' to necessity: stakeholders' involvement in place branding. *Journal of Place Management and Development* 5(1), 7–19. DOI: 10.1108/17538331211209013.
Kavaratzis, M. and Ashworth, G. (2015) Hijacking culture: the disconnection between place culture and place brands. *Town Planning Review* 86(2), 155–176. DOI: 10.3828/tpr.2015.10.
Kavaratzis, M. and Hatch, M.J. (2013) The dynamics of place brands: an identity-based approach to place branding theory. *Marketing Theory* 13(1), 69–86. DOI: 10.1177/1470593112467268.
Kavaratzis, M., Giovanardi, M. and Lichrou, M. (2017) *Inclusive Place Branding: Critical Perspectives on Theory and Practice*. Routledge, London. DOI: 10.4324/9781315620350.
Kuvan, Y. (2010) Mass tourism development and deforestation in Turkey. *Anatolia* 21(1), 155–168. DOI: 10.1080/13032917.2010.9687096.
Liu, Z. (2003) Sustainable tourism development: a critique. *Journal of Sustainable Tourism* 11(6), 459–475. DOI: 10.1080/09669580308667216.
Macionis, N. (1998) Wine and food tourism in the Australian Capital Territory: exploring the links. *International Journal of Wine Marketing* 10(3), 5–22. DOI: 10.1108/eb008683.
Maheshwari, V., Vandewalle, I. and Bamber, D. (2011) Place branding's role in sustainable development. *Journal of Place Management and Development* 4(2), 198–213. DOI: 10.1108/17538331111153188.
MapSVG (2021) Turkey map. Available at: https://mapsvg.com/maps/turkey (accessed 6 December 2021).
Marine-Roig, E. and Ferrer-Rosell, B. (2018) Measuring the gap between projected and perceived destination images of Catalonia using compositional analysis. *Tourism Management* 68, 236–249. DOI: 10.1016/j.tourman.2018.03.020.
McCracken, G. (1988) *The Long Interview*. SAGE Publications, Newbury Park, California. DOI: 10.4135/9781412986229.
McKercher, B. (2002) Towards a classification of cultural tourists. *International Journal of Tourism Research* 4(1), 29–38. DOI: 10.1002/jtr.346.

Ministry of Culture and Tourism (2007) Tourism strategy of Turkey – 2023. Available at: https://teftis.ktb.gov.tr/TR-14577/turkiye-turizm-stratejisi-ve-eylem-plani.html (accessed 12 October 2020).

Mitropoulou, A. and Spilanis, I. (2020) Towards a contemporary research agenda for island branding: developments, challenges, and dynamics. *Place Branding and Public Diplomacy* 16(4), 293–303. DOI: 10.1057/s41254-020-00181-4.

Montella, M.M. (2017) Wine tourism and sustainability: a review. *Sustainability* 9(1), 113. DOI: 10.3390/su9010113.

Morgan, N., Pritchard, A. and Piggott, R. (2002) New Zealand, 100% pure. The creation of a powerful niche destination brand. *Journal of Brand Management* 9(4), 335–354. DOI: 10.1057/palgrave.bm.2540082.

Neal, Z.P. (2006) Culinary deserts, gastronomic oases: a classification of US cities. *Urban Studies* 43(1), 1–21. DOI: 10.1080/00420980500388728.

Nocca, F. (2017) The role of cultural heritage in sustainable development: multidimensional indicators as decision-making tool. *Sustainability* 9(10), 1882. DOI: 10.3390/su9101882.

Nurse, K. (2006) Culture as the fourth pillar of sustainable development. *Small States: Economic Review and Basic Statistics* 11, 28–40. DOI: 10.14217/smalst-2007-3-en.

Official Website of Bozcaada (n.d) Nüfus durumu. Available at: www.bozcaada.gov.tr/nufus-durumu (accessed 6 December 2021).

Okumuş, D. (2018) Rethinking rural transitions: the case of Bozcaada, Turkey. PhD thesis, Newcastle University, Newcastle upon Tyne, UK.

Okumus, F., Avci, U., Kilic, I. and Walls, A.R. (2012) Cultural tourism in Turkey: a missed opportunity. *Journal of Hospitality Marketing & Management* 21(6), 638–658. DOI: 10.1080/19368623.2012.627231.

Poitras, L. and Donald, G. (2006) Sustainable wine tourism: the host community perspective. *Journal of Sustainable Tourism* 14(5), 425–448. DOI: 10.2167/jost587.0.

Prayag, G. and Ryan, C. (2012) Antecedents of tourists' loyalty to Mauritius: the role and influence of destination image, place attachment, personal involvement, and satisfaction. *Journal of Travel Research* 51(3), 342–356. DOI: 10.1177/0047287511410321.

Rebelo, C., Mehmood, A. and Marsden, T. (2020) Co-created visual narratives and inclusive place branding: a socially responsible approach to residents' participation and engagement. *Sustainability Science* 15(2), 423–435. DOI: 10.1007/s11625-019-00760-2.

Republic of Turkey (2019) 11th Development Plan (2019–2023). Available at: https://www.sbb.gov.tr/wp-content/uploads/2019/07/OnbirinciKalkinmaPlani.pdf (accessed 8 December 2020).

Ritchie, J., Lewis, J., Nicholls, C.M. and Ormston, R. (2014) *Qualitative Research Practice: A Guide for Social Science Students and Researchers*, 2nd edn. SAGE Publications, London.

Robinson, O.C. (2014) Sampling in interview-based qualitative research: a theoretical and practical guide. *Qualitative Research in Psychology* 11(1), 25–41. DOI: 10.1080/14780887.2013.801543.

Seckelmann, A. (2002) Domestic tourism – a chance for regional development in Turkey? *Tourism Management* 23(1), 85–92. DOI: 10.1016/S0261-5177(01)00066-8.

Sharma, G.D., Thomas, A. and Paul, J. (2021) Reviving tourism industry post-COVID-19: a resilience-based framework. *Tourism Management Perspectives* 37, 100786. DOI: 10.1016/j.tmp.2020.100786.

Skinner, A.M. (2002) Napa Valley, California: a model of wine region development. In: Hall, C.M., Sharples, L., Cambourne, B. and Macionis, N. (eds) *Wine Tourism Around the World: Development, Management and Markets*. Elsevier, Oxford, UK, pp. 283–296.

Smith, A. (2005) Reimaging the city: the value of sport initiatives. *Annals of Tourism Research* 32(1), 217–236. DOI: 10.1016/j.annals.2004.07.007.

Smith, D.P. and Phillips, D.A. (2001) Socio-cultural representations of greentrified Pennine rurality. *Journal of Rural Studies* 17(4), 457–469. DOI: 10.1016/S0743-0167(01)00014-6.

Spalding, M., Burke, L. and Fyall, A. (2021) Covid-19: implications for nature and tourism. *Anatolia: An International Journal of Tourism and Hospitality Research* 32(1), 126–127. DOI: 10.1080/13032917.2020.1791524.

Thode, S.F. and Maskulka, J.M. (1998) Place-based marketing strategies, brand equity and vineyard valuation. *Journal of Product & Brand Management* 7(5), 379–399. DOI: 10.1108/10610429810237673.

Tomljenović, R. (2009) Life-cycle stages in wine tourism development: a comparison of wine regions in Croatia. *Tourism Review International* 13(1), 31–49. DOI: 10.3727/154427209789130666.

Tosun, C. (2001) Challenges of sustainable tourism development in the developing world: the case of Turkey. *Tourism Management* 22(3), 289–303. DOI: 10.1016/S0261-5177(00)00060-1.

Tosun, C., Timothy, D.J. and Öztürk, Y. (2003) Tourism growth, national development and regional inequality in Turkey. *Journal of Sustainable Tourism* 11(2–3), 133–161. DOI: 10.1080/09669580308667200.
TUIK (2015) Databases. Available at: www.turkstat.gov.tr (accessed 12 October 2020).
Türker, N. and Alaeddinoğlu, F. (2016) From wine production to wine tourism experience: the case of Anatolia, Turkey. *Journal of Tourism and Gastronomy Studies* 4(3), 25–25. DOI: 10.21325/jotags.2016.40.
United Nations (2012) Future we want: outcome document of the Rio+20 United Nations Conference on Sustainable Development. Available at: www.sustainabledevelopment.un.org/futurewewant.html (accessed 6 December 2021).
United Nations (2016) Draft outcome document of United Nations Conference on Housing and Sustainable Urban Development (Habitat III). Available at: https://digitallibrary.un.org/record/1290312?ln=en (accessed 6 December 2021).
Wheeler, F., Frost, W. and Weiler, B. (2011) Destination brand identity, values, and community: a case study from rural Victoria, Australia. *Journal of Travel & Tourism Marketing* 28(1), 13–26. DOI: 10.1080/10548408.2011.535441.
Williams, P. (2001a) Positioning wine tourism destinations: an image analysis. *International Journal of Wine Marketing* 13(3), 42–58. DOI: 10.1108/eb008726.
Williams, P. (2001b) The evolving images of wine tourism destinations. *Tourism Recreation Research* 26(2), 3–10. DOI: 10.1080/02508281.2001.11081338.
World Commission on Environment and Development (1987) *Our Common Future*. Oxford University Press, Oxford, UK.
Worldometers (2020) Turkey population. Available at: www.worldometers.info/world-population/turkey-population/ (accessed 12 October 2020).
Yu, C.-P., Chancellor, H.C. and Cole, S.T. (2009) Measuring residents' attitudes toward sustainable tourism: a reexamination of the sustainable tourism attitude scale. *Journal of Travel Research* 50(1), 57–63. DOI: 10.1177/0047287509353189.
Zaichkowsky, J.L. (1985) Measuring the involvement construct. *Journal of Consumer Research* 12(3), 341. DOI: 10.1086/208520.
Zenker, S. and Braun, E. (2010) The place brand centre – a conceptual approach for the brand management of places. In: *Proceedings of 39th European Marketing Academy Conference*, Copenhagen, Denmark, pp. 1–8.
Zenker, S. and Erfgen, C. (2014) Let them do the work: a participatory place branding approach. *Journal of Place Management and Development* 7(3), 225–234. DOI: 10.1108/JPMD-06-2013-0016.
Zhou, Q.B., Zhang, J., Zhang, H. and Li, X.R. (2018) Is all authenticity accepted by tourists and residents? The concept, dimensions and formation mechanism of negative authenticity. *Tourism Management* 67, 59–70. DOI: 10.1016/j.tourman.2017.12.024.
Zouganeli, S., Trihas, N., Antonaki, M. and Kladou, S. (2012) Aspects of sustainability in the destination branding process: a bottom-up approach. *Journal of Hospitality Marketing & Management* 21(7), 739–757. DOI: 10.1080/19368623.2012.624299.

8 Cross-Cultural Understanding of the Global Multicultural Tourism Workforce: The Case of Israel (Eilat) and Jordan (Aqaba)

Ronen Shay*
Ben-Gurion University of the Negev, Beersheba, Israel; Kinneret Academic College in the Jordan Valley, Zemach, Israel

8.1 Introduction

The peace agreement between Israel and Jordan signed on 26 October 1994 (Knesset Israel, 1994) was built on the notion of peace-building instead of conflict resolution or management. Decades of hostility, suspicion and resentment accumulated over years of conflict have left scars that prevent mental and consciousness-based transition from war to peace (Sallam and Winter, 2017). Hence, this chapter aims to shed light on the immense impact that cultural aspects have on people's perceptions, attitudes and beliefs. Furthermore, it stresses the significance of cross-cultural understanding of the contemporary global multicultural tourism workforce by highlighting a single case study that could cultivate cooperation and revive the tourism industry in southern Israel.

This study focuses on a recent endeavour that signifies a new era in Israel and Jordan's relations. The agreement between the two countries signed in 2014 permits Jordanians to cross the border to work in the hospitality industry in Eilat (referred to as the 'Eilat project'; Population and Immigration Authority, 2021). The project is the upshot of the US–Jordanian Free Trade Agreement validated and implemented between 1996 and 2010. Throughout the years, 13 Qualifying Industrial Zones (QIZs) were formed, providing tens of thousands of Jordanians with employment (Sütalan, 2015). The Eilat project was an additional step to employ up to 2000 Jordanians in the hospitality industry in Israel's Red Sea resort city of Eilat amid an increasing shortage of workers, especially at the start of the high summer season. This new foundation for international business was a joint effort of the Ministry of Interior Affairs, the Israeli Hotel Association, the Eilat Municipality, the Israeli tourism department, the Aqaba Municipality, the Jordanian tourism department, human resources agencies and regular citizens (Population and Immigration Authority, 2021).

One cannot deny that security issues and bureaucracy will always be an integral part of the initiative. For the foreseeable future, monitoring, screening and inspections are vital. However, security and peace are ensured by physical and diplomatic arrangements and by engraving and anchoring the notion of peace in people's minds, hearts and souls. Hence,

*rsay@post.bgu.ac.il, ronen.fm102@gmail.com

DOI: 10.1079/9781789249149.0008

mutual economic interests have enabled the continuation of the project despite political disagreements or security hazards. The chapter aims to examine the Eilat project's national, cultural, security and managerial aspects since they constitute a profound part of Israel and Jordan's relations.

Despite the Eliat project's limitations, this chapter argues that the incorporation of Jordanian workers into the hotel industry in Eilat presents a window of opportunity to start a new phase in Israel and Jordan's relations. This business endeavour brings prospects for warmer ties between Jordanians and Israelis and adds a civilian dimension to the governmental coordination between the countries. Hence, the case study highlights that Jordan and Israel's interaction can progressively improve and expand via direct and daily connections between both countries' civilians and ultimately contribute to tourism development and planning.

The chapter is organized as follows. First it begins with the literature review, theories and approaches that constitute the base of this study. This is followed by the case study, and then the research design, describing the research aims, questions, paradigm, methodology and tools. The next section exhibits the research results generated from the research questions, while the section following that discusses the findings emerging from the research questions. The final section contains the conclusions and recommendations.

8.2 Literature Review

The study is based on three theoretical pillars: cross-cultural issues; international business management; and cross-border management. Cross-cultural issues, the first pillar, constitute the study's axis since they clarify multicultural issues in a diversified workforce. They highlight the difficulties and misunderstandings that might arise due to intercultural communication, cultural differences, language barriers, and diverse customs and norms. The relations between the Jordanian employees and their Israeli co-workers are likely to face cultural differences. Hence cultural and cross-cultural issues are the core of the study, and cultural diversity is at the heart of the interactions between the parties involved. The second pillar, international business management, delves into the human resource aspect. It encompasses planning, categorizing, organizing, leading and supervising employees and various resources to realize organizational goals across multicultural and multinational boundaries. This pillar illuminates the basic notions that generated the strategy of bringing Jordanian workers into the workforce. The human resource aspect was responsible for incorporating Jordanian workers into Eilat's hotel industry and hiring the right people for the job. Finally, the third pillar, cross-border management, stresses that border policy modifications result from negotiation between stakeholders, who share mutual interests, viewpoints and values. Border control is the outcome of collaborative management, both domestic and international.

8.2.1 Cross-cultural issues within a multicultural workforce

A multicultural workforce is gradually becoming the norm and the standard. The 21st-century corporate world is multifaceted; it is global, intricate, competitive, aggressive and dynamic. External factors (e.g. demographics, globalization, civilization and technology), as well as internal factors (e.g. competition, expectations, aspiration and accountability), are constantly impacting and shaping the operational and strategic planning of corporate management (Burke and Ng, 2006; Spinks and Moore, 2007). Hence, the section's objective is to explain how variations in the external environment impact the workforce's nature and highlight the difficulties and the potential prospects of hiring a culturally diversified staff within the Israeli tourism industry. The following paragraphs will discuss and depict the phenomenon's factors, which cause misunderstandings and obstruction in the diversified workforce.

Terms such as 'multiculturalism' and 'interculturalism' emphasize the diversity in society regarding language, religion, race and tradition. They stress how to comprehend, cope with and react to the challenges related to cultural and religious diversity. In general, modern countries are systematically unified around the dominant groups' linguistic and cultural customs. Hence, cultural minority groups struggle and encounter difficulties practising their social norms (Levey, 2012; Song, 2020). Multiculturalism and interculturalism strive to change dominant representation and communication patterns, which ostracize and marginalize certain groups. Ultimately, they encourage genuine dialogue, offer an adequate set of principles to accommodate cultural diversity and underline the significance of communication. These terms are not limited to identity struggles and cultural recognition. They also focus on economic interests and political power and encompass claims to resolve and remedy financial and political shortcomings that disadvantage people from marginalized groups (Levey, 2012; Song, 2020). Since Arabs are a marginalized group in Israel, it is of interest to view if the discourse with the Jordanian workers is affected by inherent identity and cultural struggles.

Geographically, Israel and Jordan are neighbouring countries. Historically, they are arch enemies who have found a way to live peacefully side by side. But culturally, they are pretty different in terms of norms, values, manners, behavioural patterns, rules, philosophies and policies. The peace treaty between Israel and Jordan reflects aspects of normalization (Sütalan, 2015). The countries established mechanisms for neighbourly relations and formed foundations for economic collaboration, tourism, cultural and scientific exchange, transportation, civil aviation, environment, energy, health and agriculture collaboration. Unfortunately, the Jordanian–Israeli peace process has, in practice, failed to produce good or friendly relations. The communication between the countries continues to be primarily intergovernmental. Although the agreement achieved diplomatic recognition, diplomatic relations are still chilly. The trade between Jordan and Israel remains modest and limited, bilateral trade and tourism did not yield the anticipated outcomes, and despite growth in mutual commerce, it is not a constant or stable increase. The coldness between the nations, which manifests in tensions and confrontations, is mainly due to the effect of the unresolved Palestinian–Israeli conflict, the controversy surrounding east Jerusalem and the reluctance of the Jordanian people to normalize the relationship (Sütalan, 2015). Hence, the decision to integrate Jordanian workers into the tourism industry in Eilat necessitates a close examination of the difficulties and misunderstandings that might arise due to intercultural communication (i.e. interaction between people of varied cultures), cultural differences, language barriers and various norms.

Over the last few decades, we have witnessed a striking transformation in the workforce's nature and how we work. Today's labour force is progressively heterogeneous since it comprises diverse workers (Mathis and Jackson, 2004) in terms of gender, age and culture. For example, features, ethics, proficiencies, manners, habits, behaviours, demands and expectations of the new and younger workforce (i.e. Generation Y) are notably distinct from preceding generations (Burke and Ng, 2006; Spinks and Moore, 2007). This study illuminates generational differences by emphasizing the younger generation's reluctance to do manual work (Construct-Ed, 2018). This reluctance, along with the high level of hospitality employee turnover (e.g. Salary.com, 2018), obligated the hotel industry to find a solution across the Israeli border. Since the relations between Jordanian employees and their Israeli co-workers are likely to face cultural differences, cultural and cross-cultural issues lie at this study's heart.

Globalization is the hallmark of today's economy. It has formed and generated a global marketplace for goods and services, interlinked local markets, reduced information transmission costs, and removed political and cultural barriers. Moreover, it has motivated the global labour market, international cooperation, globally distributed work teams and stimulated competition for talent (Burke and Ng, 2006). A prominent characteristic and consequence of globalization is outsourcing, offshoring and contracting out services. Once companies understood that they could cut costs by transferring work to countries where labour can be acquired inexpensively, globally distributed teams comprising workers from various countries, different time zones and diverse work practices started to form (Burke and Ng, 2006; Spinks and Moore, 2007). An additional significant outcome of globalization is the change in the employment relationship. In contrast to the past, contingent workers signify a growing percentage of employees in today's workforce since work is being parcelled out to independent contractors and consultants (Burke and Ng, 2006). Hence, today's intricate marketplace requires managers to quickly comprehend how globalization, civilization, technology, generational and cultural diversity, and the political, social and economic climate impact policies and procedures (Spinks and Moore, 2007).

The fast pace of internationalization and globalization carries requisites, constraints and obligations for the unavoidable and inescapable integration of cultures. As a concept and in actuality, culture is all-embracing and multidimensional. It impacts our identity and the way we interconnect within and across regional and national borders. Culture gives a sense of distinctiveness and belonging; it is a set of values, attitudes, concepts, norms and beliefs that dictate learned behaviours shared within a specific society. It awards a feeling of harmony, constancy, unity, predictability and certainty. Thus, employees reflect their respective cultures in domestic and global workplace settings (Condruz-Băcescu, 2010; Kawar, 2012).

Cross-cultural challenges in international business management have become a prominent research topic (Bhatt, 2014). The modern world is a global village getting smaller and smaller, thus summons new challenges that stem from the world's varied cultural and ethnic landscape. National culture has continuously been a complicated and challenging concept to define. Cultures are nuanced, and in today's globalized world where they connect and collide, the art of finding distinct features is becoming even more challenging (Kawar, 2012; Swarnalatha and Tephillah Vasantham, 2014). Generally, five trends in culture have an immense effect on the marketplace: (i) enhanced awareness of cultural disparities in domestic and global workplaces; (ii) expanding need for cross-cultural understanding; (iii) managing talent internationally; (iv) bigger and better emphasis on global leadership competencies; and (v) increased usage of virtual global teams' human research.

It is not hard to comprehend how and why cultural distinctions cause different insights and misunderstandings and quickly create an obstruction for multinational companies. Corporate culture has typical characteristics like norms, behavioural regularities, values, symbols, rules, philosophies, policies and organizational climate (Singh and Kaur, 2016). Therefore, working with people from different cultures necessitates intercultural communication and cultural diversities like conduct, communication, and proper conversation and interaction (Jandt, 2001; Kawar, 2012). Given these points, it is simple to understand why accidents of miscommunication occur. Hence, it is interesting to explore the factors involved in how the hotel industry workers manage communication with the Jordanian workers in Eilat.

Numerous factors hinder the successful implementation of intercultural communication and the construction of a global mindset. First, one of the main challenges of managing a multicultural environment is 'culture shock', namely the anxiety, insecurity and confusion caused by encountering another culture. Lack of knowledge of how to act generates a feeling of disorientation and isolation, ultimately motivating a return to the homeland. A reverse challenge is parochialism, in which people fail to identify and acknowledge critical variances between their own culture and other cultures or regard these differences as irrelevant or unimportant. An extension of parochialism is prejudice and ethnocentrism, which transpire when people are convinced that their homeland terms and settings are idyllic. Therefore, they are unwilling to accept different social systems or, for instance, judge the conditions in a new country by the standards of their homeland. The problem arises when each participating party is sure that their operation is the right way (Condruz-Băcescu, 2010; Singh and Kaur, 2016).

The contradictory cultural dimensions generate misunderstanding, mistrust and miscommunication between individuals who originate from diverse backgrounds. Hence, hoteliers must comprehend and adapt sensitively to various workers' wants and needs and strive to keep this gap to a minimum (Swarnalatha and Tephillah Vasantham, 2014). The starting point for successfully managing intercultural standards is recognizing the validity and legitimacy of an individual's cultural patterns and realizing the cultural base (Condruz-Băcescu, 2010). Thus, for the hotel industry to thrive, it needs managers who comprehend and appreciate a culturally diverse workforce and work proficiently with diverse cultures that have varying norms, values, work ethics and business protocols. As a result, intercultural communication and cross-cultural intelligence are necessary for tourism. Namely, the capacity to switch from ethnic and/ or national settings and swiftly acquire new social interaction patterns and suitable behavioural reactions is obligatory.

Roberts *et al.* (1998) detected three critical challenges in managing a global workforce. First, deployment, namely, allocating the skills where they are needed irrespective of geographical location. Second, knowledge and innovation dissemination, that is to say, dispersal of state-of-the-art knowledge and practices regardless of where they originate. Third, talent identification and development, namely, finding and advancing candidates with the right skills to succeed in multicultural environments. Under these circumstances, forming an ideal business relationship entails global fluency, defined as a facility with cultural behaviours that help a chain flourish in an ever-changing global business environment (Tosti, 2002). In other words, intercultural communication and cross-cultural understanding are prerequisites in establishing and maintaining a good business relationship. Intercultural communication is the optimal way to help employees become more well-informed about cross-cultural communication in terms of their cultural standards, manners, assumptions and etiquette, as well as those of other cultures.

Furthermore, technological advances and globalization have transformed the nature of the workplace. Boundaries no longer limit it, and it now has to confront challenges associated with cultural aspects (Swarnalatha and Tephillah Vasantham, 2014). Given these points, the key to overcoming contemporary organizational challenges is a successful implementation of diversified, cross-cultural management. Globalization is no longer exclusively assessed on the basis of economic growth; instead, it integrates social, cultural, political and historical progression. In a globalized world in which, on the one hand, distances have diminished, and on the other hand, relationship intricacy has intensified, the workplace's multiculturalism should be considered a fixed and strategic variable of organizational management (Celano Teixeira *et al.*, 2015).

Hotel chains believe in diversity as it assists them in tackling different problems. Individuals from varied cultures bring distinctive skill sets to the table. Therefore, the optimal way to integrate the diverse workforce is by forming solid relationships through personal bonds and mutual understanding (Swarnalatha and Tephillah Vasantham, 2014). In conclusion, to realize organizational goals and evade potential risks, the tourism and hospitality industry should be culturally sensitive and advance teamwork, creativity, and motivation through flexible leadership.

8.2.2 International business management

The aforementioned difficulties of high employee turnover and Generation Y's reluctance to work in cleaning jobs faced by resorts in Israel sparkled the 'Eliat Project' initiative. International business management encompasses planning, categorizing, organizing leading and supervising employees and various resources to realize organizational goals across multicultural and multinational boundaries. The field of international management offers possibilities and challenges to managers, businesses and governments. Nowadays, due to global market expansion, managers ought to be familiar with international trade and finance traits such as culture, political structure, foreign exchange, geographical terrain, time, food and technology. Global management involves crossing cultures, multinational corporations' relations, universal viewpoints and corporate issues. In the globalized world, international managers must acknowledge and respect various ideas, values, beliefs and traditions, while guaranteeing that the allocation of human resources attains the organization's goals (JNU, 2013).

In the past, human resource managers and human capital strategies were seen as part of the administrative side of an organization. However, nowadays, this perspective has shifted, and human capital management is viewed as one of the chief considerations in business strategies (Dash, 2012). Human resource management is a dominant facet in developing and advancing the skills, attitudes, knowledge and capabilities of a successful organization. Human resource management helps foster a culture in which clever strategies are envisioned and revised or bad choices are hindered before their implementation (Boxall and Purcell, 2000). Human capital management is about safeguarding the vast potential provided by people while ensuring its alignment with the business's mission and strategic objectives (Finn, 2003). Thus, the fruitful application of knowledge management initiatives is reliant on the effective management of human resources. In this study, the human resource aspect was ultimately responsible for incorporating Jordanian workers into the hotel industry in Eilat and hiring the right people for the job.

8.2.3 Cross-border management

International borders are dynamic across time and space. They form, are removed and change via various material and social mechanisms. Disagreements, commerce, incorporation and omission impact how borders between neighbouring states are regulated (Hataley and Leuprecht, 2018). Borders are institutions with specific rules and behaviours that shape their functionality, preservation and modification (Newman, 2003). They influence the way people think of frontiers, while their role and evolution mirror states' values and interests. However, borders are human constructs generated to control human behaviour in a spatial context (Popescu, 2012).

Border management forms an equilibrium between security and the legal flow of people and goods. This notion is based on the comprehension that there is conflict between the need to avert unwelcome people and unwanted goods from crossing borders and the need to boost the state's economic strength via trade and travel. Thus, cross-border management offers opportunities to enhance security while concurrently improving the flow of legal passengers and goods through policies, programmes, regulations and activities (Gerstein *et al.*, 2018). Border management is grounded on three principles that maintain the delicate balance between security and economic strength. The first notion is integrating safety and the flow of people and trade. The second is creating an integrated, layered approach, namely, developing a comprehensive system that includes planning, equipment and training. Finally, forming a border management system necessitates unity of effort across all related governmental and non-governmental stakeholders – notably, interested parties with responsibilities or gains in national borders. Therefore, command, control and communication that admit collaboration and synchronization among numerous interagency organizations and from the national to the local level are vital to border control (Gerstein *et al.*, 2018).

Modifications in border policy result from negotiation between stakeholders, who share mutual interests, viewpoints and values; the procedure is not exclusive to elites (e.g. politicians, legislators and people of means), nor is it linear. Instead, it is a variable with multiple non-state actors (e.g. business people, administrators and entrepreneurs) whose varied social insights are mirrored in the outcome. Thus, borders mirror the historical, political and cultural framework of a state. They echo societies and cultures, and encompass and reflect specific social relationships in space and time (Hataley and Leuprecht, 2018). Hence, the handling of border control is the responsibility of both sides of the border. It has to involve collaborative management, including domestic (i.e. intra-service and inter-agency cooperation) and international dimensions (i.e. transnational collaboration) (Rusdiyant, 2017).

8.3 Case Study

Eilat is a prominent resort city in Israel, and during summer vacations, Israelis and international tourists visit the city for recreational activities, attractions, the desert and the Red Sea. However, the hospitality industry in Eilat suffers from a lack of employees, mainly in housekeeping and dishwashing positions. The labour shortage influences the service quality of the hotels. Hence, the peace agreement between Israel and Jordan signed in 1994 (Knesset Israel, 1994) initiated a pioneering solution for the labour shortage (Population and Immigration Authority, 2021). This novel solution is the focal point of this chapter.

The pact between the countries was more than a formalization of an actual ceasefire. In essence, it profoundly altered the nature of the Israeli–Jordanian relationship, increasing safety, stability and US interests in a chaotic region (Sütalan, 2015). However, the expectation that the peace agreement would lead to warm and neighbourly relations based on coexistence, mutual cooperative endeavours and an acceptance of the other has failed to materialize. Israel was unable to stimulate the governmental system to instigate economic collaboration, proposals and projects. Concurrently, the internalization of the peace by the Jordanian people did not happen, and the bureaucracy exhibited suspicions towards Israel (Zalzberg, 2020). Therefore, in practice, despite hopes and aspirations, the peace between the nations has remained 'cold' (Sütalan, 2015; Sallam and Winter, 2017).

As stated earlier, the Eilat project is a venture that denotes a new chapter in Israel and Jordan's relations. The agreement between the 2 countries permits 2000 Jordanians to cross the border to work cleaning jobs in the hospitality industry in Eilat. Nevertheless, the project is strictly regulated. For instance, it compels the Jordanian workers to return to Jordan by 11pm, prevents them from leaving the Eilat city perimeters and restricts them to cleaning jobs (Population and Immigration Authority, 2021). Thus, the research examines the opportunities and shortcomings of implementing the project model due to the obstacles and merits.

8.4 Research Design

The study focuses on understanding a phenomenon, integrating Jordanian workers in the hospitality industry, as a humanistic approach. The objective is to capture the delicate social reality of Israelis and Jordanians and understand their standpoints, notions, beliefs, thoughts, experiences, behaviour and interactions. Hence, a qualitative research approach was implemented to study the perception of Israeli and Jordanian officials and Jordanian labourers working in Eilat about the Eilat project. The research tools included interviews, focus groups, observations and document analysis, which enabled the study of standpoints, insights, and assessments of the various officials and stakeholders involved in the project.

8.4.1 Research aims and questions

The research aims to identify the merits of the Eilat project and its contribution to tourism and the hotel industry. It examines the opportunities and shortcomings of implementing the project model and investigates its replicability in other regions and countries. Three micro-level goals accompany the chief objective: (i) to ascertain intercultural business management issues between Jordanian and Israeli managers; (ii) to investigate how the Jordanian employees and the Israeli managers are coping with cross-cultural issues, i.e. religion, mentality and quality control points of view; and (iii) to observe the entrepreneurial aspects of implementing the project. Ultimately, the study strives to answer two fundamental questions:

1. What are the opportunities and shortcomings of implementing the project model, recruiting culturally diverse employees from across the border to work in the hospitality industry?
2. What is the feasibility of replicating the model to other countries? Is it possible to identify common ground between the EU and the Israel–Jordan Eilat project?

8.4.2 Research population and sampling

The research population in the study shares similarities. Namely, all the participants are directly or indirectly connected to the hotel industry or the decision to incorporate Jordanians into Eilat's hospitality industry. The research population includes 38 participants from 4 fields and different professions: politics (e.g. mayors), legislation (e.g. member of the Knesset), business (e.g. CEOs, entrepreneurs and business people) and administration (e.g. hotel managers, administrators, human resource managers, coordinators and line employees). The participants differ in terms of age, gender, work experience, education and place of residence (i.e. participants from both sides of the border). Thus, the participants' similarities and variances provided a comprehensive perspective, inclusive insights and varied viewpoints.

8.4.3 Research tools

Twenty face-to-face interviews were conducted with diversified people from different lines of work. Questionnaires were prepared to address the participants' occupation and their relevance to Jordanian workers' integration into Eilat's hotel industry. We assigned the interview formats to human resource managers, hotel managers, housekeeping managers (i.e. administrators), business people and decision makers (i.e. politicians, legislators). The interviews included an assortment of questions that aimed to uncover the respondents' stances and attitudes towards integrating the Jordanian workers and questions that tried to comprehend the effect of economic, cultural, political and security factors on the success or failure of this endeavour. Furthermore, three focus groups took place: two groups encompassed five people, and one group included eight people. One focus group consisted of Jordanian workers, and the other two contained business people from Eilat and Aqaba. The focus groups were directed by a moderator's guide, which included engagement, exploration and exit questions. We asked the Jordanians about their concerns, the screening process, their level of satisfaction, collegial feelings and the impact of financial aspects. The interviews with business people were about security issues, the project's effect on the Israeli employment market, cultural diversity, challenges and political implications.

Observations grant researchers the opportunity to look for non-verbal expressions of emotions, uncover who interacts with whom and understand how participants connect. In this study, the researcher took the stance of the observer as participant, which enabled him to take part in the group activities while preserving his primary role of collecting data. In this stance, the researcher was an observer who is not a member of the group and interested in participating in conducting better observations and, thus, generating a complete understanding of the group's activities. It is noteworthy that the group studied was aware of the researcher's observation activities. During observations, the researcher focused on the interaction and relationship between Israeli and Jordanian workers, the social and professional incorporation of Jordanian workers in the workplace and tourists' attitudes towards the new employees.

We also researched and examined various documents such as press releases, programme proposals, survey data, organizational and institutional reports, public records, immigration policies, government decisions, labour and employment laws, administrative guidelines, municipal plans, regional forum statements and bilateral agreements. These documents provided the necessary information about the context and background that enabled the incorporation of Jordanian workers into the hotel industry and strengthened the research's scientific method, credibility and reliability. Table 8.1 presents the research population profile.

8.4.4 Data analysis

The case's depth and scope were studied from diverse angles such as cultural, political, financial and occupational. Therefore, the theories and methodic approaches yielded five themes (cultural aspects, Israel and Jordan relations, interpersonal interactions, human resource management and entrepreneurship) related to the research questions. Each theme instigated categories that portray the essence of the general topics. Three issues – (i) cultural aspects; (ii) Israel and Jordan's relations; and (iii) interpersonal relations – relate to the first research question: what are the opportunities and the shortcomings entailed by implementing the project model, recruiting culturally diverse employees from across the border to work in the hospitality industry? The fourth and fifth subjects, human resource management and entrepreneurship, refer to the second research question: what is the feasibility of replicating the model in other countries?

Table 8.1. Research population profile.

	Job description[a]	Years of experience
Politics		
	Regional liaison	15
	Mayor	17
	Initiator of the project	20
Business		
	Business people from Eilat and Aqaba (10)	5–20
	Acting director for investment promotion	12
	Division head, Marketing & Promotion (2)	5
Administration		
	Member of the Knesset	15
	General manager of Eilat Hotel Association	13
	Head of administration population of Eilat	40
	Chief of HR hotel chain (3)	25–40
	General manager of a hotel (5)	8–25
	Manager of an employment centre (hotel chain)	10
	Jordanian human resources company representative	10
	Manager of the largest Jordanian HR company	20
	Jordanian employees (8)	2–10

[a] Numbers in parenthesis = the number of interviewees.

8.5 Findings

As stated, the interviews' review and analysis generated five themes, and each theme instigated categories that portray the essence of the general topics.

8.5.1 Cultural aspects

Examination of the data revealed that the first theme, cultural aspects, could be categorized into three motifs: language; religion and political affiliation; and ethnicity and race. First, language barriers hinder communication and inhibit social interaction. However, willingness to teach, learn, speak a foreign language, implement technological tools or simple human gestures can bridge linguistic differences and gaps. Second, the results signify that ethnicity is detrimental to trust and confidence in racially different people. With that said, close encounters, positive attitudes and trust-building steps can change people's instinctive perceptions. Third, the results indicate that mundane activities like work can be kept separate from religion and political affiliation. Namely, people decide whether to instigate religious or political discussions. These debates are not inevitable and can be avoided to maintain amicable work relations.

8.5.2 Israel and Jordan's relations

The second theme, Israel and Jordan's relations, emphasizes the complex ties between the countries. It portrays the implications of cold peace while illustrating the merits of the project that constitute a beacon of hope. Inspection of the data disclosed that the theme could be categorized into three motifs: security issues and fear of terror attacks, bureaucracy, nationality and citizenship. The results point out that security issues and bureaucracy will always be a fundamental part of this initiative. For the foreseeable future, monitoring, screening and inspections are crucial. However, mutual interests have enabled the continuation of the project despite political disagreements or security hazards. Furthermore, the results show that true peace is achievable through human connections. Social encounters generate trust, kinship and closeness, and eradicate prejudice and misconceptions.

8.5.3 Interpersonal relations

The third theme focuses on interpersonal and intersocial relations. It highlights the connections between Israelis and Jordanians while stressing sentiments of trust and respect, which the project managed to form. A review of the data disclosed that the 'Interpersonal relations' theme could be categorized into three motifs: building trust and credibility, respect and recognizing each other's unique characteristics. The findings illustrate that the project's highest achievement is the mutual respect and trust it could forge between two nationalities. The initiative managed to change perceptions, overcome negative views, and facilitate acknowledgement of the other side's unique characteristics.

8.5.4 Human resource management

The fourth theme delves into the recruiting, screening, employment and management of Jordanian workers. The analysis identified four categories: screening, recruitment, training, compliance with local standards, job execution and interaction with the other employees. The outcomes demonstrate that even though the recruitment and screening process is long, complex and complicated, the result is satisfactory and pleasing. The hoteliers are pleased to recruit loyal and experienced workers, and the Jordanians are delighted to find a job that pays well.

8.5.5 Entrepreneurship

Along with national, social and cultural outcomes, the initiative has economic and business effects. Hence, the fifth theme emphasizes the business opportunities and the regional upshots of the project. The findings reveal that the project has a significant monetary impact on the economies of Israel and Jordan. In addition, the initiative encourages entrepreneurship and facilitates dreams about expanding the programme and realizing it in more cities.

8.6 Conclusions and Implications

The conclusions that emerged from this study enable the advancement of an evidence-based model of a cross-cultural contemporary global multicultural workforce. Figure 8.1 demonstrates the model. The model provides a new conceptual framework of procedures to facilitate cooperation between countries that do not necessarily maintain friendly or peaceful relations. It combines two main factors: intercultural management aspects and cross-cultural aspects. When there is synergy between these aspects, a global multicultural workforce project is achievable.

The research findings clarified that different aspects hamper the progression of the project. Features such as language barriers, historical background, security issues, bureaucracy, religion, ethnicity and political affiliation put a spoke in the initiative wheel and complicate the process. Yet, the project manages to flourish despite the setbacks. Its merits exceed the difficulties and obstacles. In a short time, the project succeeded where political peace has failed. It forged friendships and business plans. Hence, all in all, the project's virtues surpass its shortcomings.

Several factual conclusions emerged from research question 1. The first theme, cultural aspects, generated three deductions about the cross-cultural understanding of the contemporary global workforce. First, a cross-cultural understanding is associated with a steady communication flow through mutual language. Second, a cross-cultural understanding is related to the cultural knowledge of the other side's ethnicity and race. Third, cross-cultural and intercultural sensitivity is facilitated by cultural and intercultural sensitivity based on

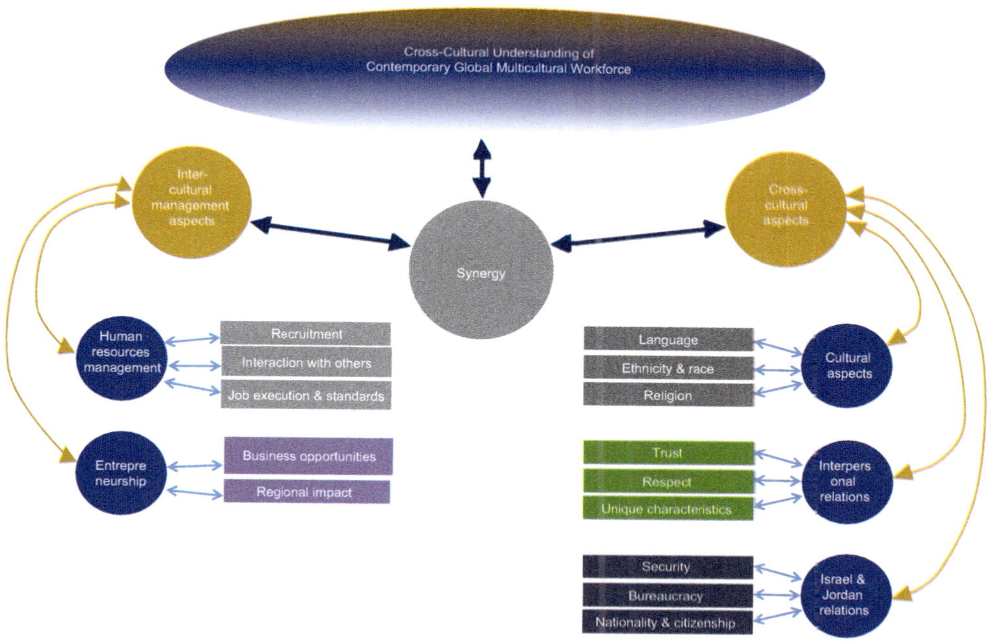

Fig. 8.1. Cross-cultural understanding of the contemporary global multicultural workforce.

mutual respect to religious customs, opinions, beliefs, political tendencies and sacred laws.

The conclusions outline the means to successful management of an intercultural workforce. They emphasize that the complex marketplace compels managers to swiftly understand how cultural diversity and the political, social and economic climate impact policies and practices (Spinks and Moore, 2007). Furthermore, deductions stress that working with people from diverse cultures requires intercultural communication and comprehension of cultural differences like conduct, communication and proper means of conversation and interaction (Kawar, 2012).

The research findings show that if hoteliers are culturally sensitive and recognize the validity and legitimacy of an individual's cultural patterns (Condruz-Băcescu, 2010), they can diminish conflicts, mistrust and disagreements. The findings align with Condruz-Băcescu and Singh's insights regarding the necessity of intercultural communication and cross-cultural intelligence. The managerial implication is that hotel managers should attain new social interaction patterns and suitable behavioural reactions when interacting with an international workforce.

The second theme, Israel and Jordan's relations, gave rise to numerous conclusions. Primarily, for the foreseeable future, security issues and bureaucracy will be entangled and inseparable from the Eilat project. Yet mutual interests and social interaction are a step in the right direction to guarantee a warmer peace and safeguard the initiative's progress while executing the current global multicultural workforce approach.

Concerning security issues and fear of terror attacks, the study detected that a cross-cultural understanding of the contemporary global multicultural workforce is characterized by a fear of terror attacks that necessitate continuous monitoring and border security checks. Furthermore, regarding bureaucracy, the analysis noted that a cross-cultural understanding is strongly associated with inevitable bureaucratic procedures that enable the project's continuation. Finally, on the subject of nationality and citizenship, the examination perceived that a cross-cultural understanding is related to mutual interest and close cross-cultural encounters that facilitate good working relations between the parties. The cold peace between Israel and Jordan is reflected in the stalemate in economic cooperation, suspicions regarding the other side's intentions and resentment which accumulated over years of conflict (Sütalan, 2015; Sallam and Winter, 2017). The coldness between the nations manifests in tensions and confrontations. Hence, it is no wonder that Israelis and Jordanians had reservations about the implementation of the project.

A lack of cross-cultural sensitivity might negatively impact communication, morale, teamwork and turnover. Therefore, Eilat's hoteliers and employees understood that they had to comprehend and appreciate a culturally diverse workforce for the endeavour to thrive and work in a setting that holds varying norms, values, views and beliefs. Both sides grasped that the best way to integrate diverse notions and faiths is by building a solid relationship through mutual understanding.

The third theme, interpersonal relations, derived various notions. The concept of building trust and credibility elicited that a cross-cultural understanding of the contemporary global multicultural workforce is characterized by fostering confidence and trustworthiness due to employers' and employees' behaviour, i.e. instilling a sense of equality, respect and commitment. Regarding the motif of respect, the findings showed that a cross-cultural understanding is instigated by providing a culture of equality among the employees about salaries, honour and consideration. Concerning recognizing each other's unique characteristics, it was evident that cross-cultural understanding can only succeed if both parties utilize the maximum benefit from their employees.

Several factors might hinder the successful implementation of intercultural communication and the construction of a global mindset (Condruz-Băcescu, 2010; Singh and Kaur, 2016). For instance, 'culture shock', narrow-mindedness, prejudice and ethnocentrism. However, the research findings illustrate that Israelis and Jordanians managed to avert these issues through understanding, consideration and respect – both sides' willingness to go the extra mile resulted in mutual respect and friendships.

Diverse factual conclusions arose from research question 2 (i.e. the feasibility of replicating the model in other countries). The fourth theme, human resource management, exhibited different ideas. On screening, recruitment and training, the study demonstrated that cross-cultural understanding of the contemporary global multicultural workforce is associated with a rigorous recruitment process under clear human resource management principles. Regarding compliance with local standards, the findings indicated that cross-cultural understanding is related to the workers' close compliance with their employers' work standards and expectations. As to job execution, the study concludes that cross-cultural understanding is characterized by a high level of job execution by professional workers fit for the job. Concerning interaction with the other employees, the analysis perceived that cross-cultural understanding is allied with intercultural communication and interaction.

As mentioned before, the starting point for successfully managing intercultural standards is recognizing the validity and legitimacy of an individual's cultural patterns and realizing your colleague's cultural base (Condruz-Băcescu, 2010). Thus, hoteliers understood that they ought to comprehend and appreciate a culturally diverse workforce for hotels to thrive. As a result, they initiated intercultural communication and swiftly learned new social interaction patterns; hence, they enabled authentic interaction between employees. In this sense, intercultural understanding and consideration can form productive collaboration and close relationships after work hours.

The fifth theme, entrepreneurship, revealed clear conclusions. First, concerning business opportunities, it is prominent that cross-cultural understanding of the contemporary global multicultural workforce is associated with business opportunities that bring about prosperity for all. The project allows further investments, enables more endeavours and helps families from both sides of the borders to build businesses and prosper. Second, regarding regional impact, it was noticeable that cross-cultural understanding is related to an atmosphere of neoliberalism where individuals are free to pursue prosperity and wealth through implementing shared economic interests, thus enhancing regional benefits.

To date, there is no definite answer to the feasibility of duplicating the project. The initiative may be implemented in other areas in Israel, copied at the Israeli–Egyptian border, or even applied in the US and Mexico. Perhaps implementing the initiative in different regions will solve unemployment issues, prevent illegal immigration, fill unattractive jobs and unwanted positions, help overcome poverty and strengthen the hotel industry. However, the situation in Europe is different. Contrary to the setting in Israel, European countries deal with immigration and refugees who want to assimilate into the community. Perhaps implementing the initiative on European soil will reduce unemployment, diminish deprivation, and reinforce tourism and hospitality.

8.6.1 Contribution to knowledge

The research has theoretical and practical contributions. Theoretically, the chapter incorporates different approaches and lines of thought into a single case study (i.e. integrating Jordanian workers into the hotel industry in Eilat) and produces an innovative model. The model is rooted in theoretical thought. It considers the importance of culture in the workplace and the significance of human capital management. Hence, the proposed model, which includes various aspects (e.g. cultural, economic and managerial), maintains the potential to impact tourism and the hospitality industry.

Practically, the study findings and conclusions can be implemented and executed in other regions in Israel or worldwide. The model helps future endeavours in the planning process. It contributes to knowledge, efficiency, finance, trust-building, screening, recruitment and training. Moreover, the model yields a reliable management tool since it considers various aspects and needs, such as culture, security, economy, human resource management and intercultural relations.

8.6.2 Practical implications and recommendations

The unique and ambitious Eilat project overcame multiple problems and difficulties. For instance, the cold peace between Israel and Jordan, cultural differences and ongoing political disputes (e.g. intifadas, confrontations and terror attacks). However, both sides conquered apprehensions and fears and agreed on processes and procedures that ensure the Israeli and Jordanian employees' security and safety. Israelis and Jordanians understood that even primordial fears could be modified and amended through close and personal encounters, gratifying gestures and pleasant experiences. They implemented an open-door policy to create unity and equality, optimal work conditions and respect.

In addition, the initiative has a positive financial effect on Jordan and Israel. Jordan could reduce its unemployment rate, while Israel could save the hospitality industry in Eilat and improve its recreational facilities. Other sectors, such as transportation, human resources, changes and housing, also profited from implementing the project. Moreover, the project benefited all the parties involved: the Jordanians earn high salaries, and hoteliers are pleased with the work level and hope to incorporate the Jordanians into other departments such as maintenance and reception. Additionally, guests enjoy clean and tidy hotels and upgraded facilities. The project's screening, recruitment and training processes proved to be effective: turnover decreased, Jordanians' integration into the workforce was successful and hoteliers could focus on other initiatives. To sum up, the project's accomplishments manifested in the initiation of more cooperative projects (e.g. aviation agreements, joint industry projects, energy trilemma and marine biology).

It is noteworthy that the COVID-19 pandemic raised concerns about the project's feasibility and generated new restrictions. Years of conflicts, disputes and clashes did not result in border closure, a reality that the spread of the virus dictated. COVID-19 is the new regulator in the newfound world; it directs and commands daily conduct and border control. Since the reopening of Israeli hotels, the project's rules have changed. A limited number of Jordanian workers were allowed to cross the border but were ordered to stay in Israel until further notice.

The study's limitations derive from the nature of the research paradigm and the decision to focus on a single case study.

Semi-structured interviews and focus groups might generate social desirability. Hence, to overcome this shortcoming, no judgements were formed during meetings to allow for candid and sincere outlooks to be collected. Furthermore, qualitative research holds high validity yet maintains a low reliability level. Therefore, to enhance confidence in the findings, we used numerous approaches and methods to validate the data and display a robust body of conclusions. Thus, generalization is restricted; nonetheless, it is applicable if comparable settings are located.

Future research should expand the research population in two aspects: size and scope. Regarding size, more people could be interviewed to validate the results. The study encompasses 38 participants; enlargement of the population will enhance reliability and generalization of the findings. The scope can also be expanded by interviewing people who are influenced by the integration of Jordanian workers, for example, the residents of Eilat, tourists, security agents and the workers' families in Jordan.

This research is qualitative; future studies can incorporate quantitative features to strengthen the study's findings and conclusions. Moreover, this chapter focused on five distinctive aspects of the Eilat project; however, different facets such as politics, travel, joint ventures, commerce and trade can similarly elaborate on the initiative's success or shortcomings.

References

Bhatt, K.T. (2014) Cross-cultural challenges in the international business management. LinkedIn. Available at: https://www.linkedin.com/pulse/cross-cultural-challenges-international-business-management-bhatt (accessed 23 April 2019).

Boxall, P. and Purcell, J. (2000) Strategic human resource management: where have we come from and where should we be going? *International Journal of Management Reviews* 2(2), 183–203. DOI: 10.1111/1468-2370.00037.

Burke, R.J. and Ng, E. (2006) The changing nature of work and organizations: implications for human resource management. *Human Resource Management Review* 16(2), 86–94. DOI: 10.1016/j.hrmr.2006.03.006.

Celano Teixeira, A.C., Lima, D.M. da C., Guedes, A.L.M. and Maciel, E. (2015) Transnational companies and multiculturalism: challenges for analysis models in international management. *Revista Pensamento Contemporâneo Em Administração* 9(2), 84. DOI: 10.12712/rpca.v9i2.544.

Condruz-Băcescu, M. (2010) Managing a multicultural environment. *Synergy* 6(1), 85–93.

Construct-Ed (2018) Millennials (still) don't want the trades. Here's why & what needs to happen ...Available at: https://www.construct-ed.com/millennials-still-dont-want-trades-heres-needs-happen (accessed 27 October 2021).

Dash, S.P. (2012) Human Capital Management (HCM): the evolution of the field. Available at: https://www.academia.edu/34509542/Human_Capital_Management_HCM_The_evolution_of_the_field (accessed 27 October 2021).

Finn, R. (2003) Human–capital management (HCM): a three-letter acronym too far for HR or the way to get the people agenda on to that of the board? *Human Resource Management International Digest* 11(5), 2–4.

Gerstein, D.M., Atler, A., Davenport, A.C., Grill, B., Kadlec, A. *et al*. (2018) *Managing International Borders*. RAND Corporation, Santa Monica, California.

Hataley, T. and Leuprecht, C. (2018) Determinants of cross-border cooperation. *Journal of Borderlands Studies* 33(3), 317–328. DOI: 10.1080/08865655.2018.1482776.

Jandt, F.E. (2001) *Intercultural Communication: an Introduction*, 3rd edn. SAGE Publications, Thousand Oaks, California.

JNU (2013) *International Business Management*. Jaipur National University, Jaipur.

Kawar, T.I. (2012) Cross-cultural differences in management. *International Journal of Business and Social Science* 3(6), 105–111.

Knesset Israel (1994) The Peace Agreement between Israel and Jordan. Available at: https://www.knesset.gov.il/process/docs/peace-jordan.htm (accessed 27 October 2021).

Levey, G.B. (2012) Interculturalism vs. multiculturalism: a distinction without a difference? *Journal of Intercultural Studies* 33(2), 217–224. DOI: 10.1080/07256868.2012.649529.

Mathis, R.L. and Jackson, J.H. (2004) *Human Resource Management*. SouthWestern, Nashville, Tennessee.

Newman, D. (2003) On borders and power: a theoretical framework. *Journal of Borderlands Studies* 18(1), 13–25. DOI: 10.1080/08865655.2003.9695598.

Popescu, G. (2012) *Bordering and Ordering in the Twenty-First Century: Understanding Borders*. Rowman & Littlefield Publishers, Lanham, Maryland.

Population and Immigration Authority (2021) General Conditions for Incorporating Jordanian Workers in the Hotel Industry in Eilat. Policy 9.6.0003. Available at: https://www.gov.il/BlobFolder/policy/jordanians_employment_eilat_hotels_terms/he/9.6.0003.pdf (accessed 25 October 2022).

Roberts, K., Kossek, E.E. and Ozeki, C. (1998) Managing the global workforce: challenges and strategies. *Academy of Management Perspectives* 12(4), 93–106. DOI: 10.5465/ame.1998.1333982.

Rusdiyant, R. (2017) Collaborative border management as the strategic revitalization of state border management in Indonesia. *International Journal of Pure and Applied Mathematics* 116(24), 415–426.

Salary.com (2018) How does turnover in your organization compare to average turnover. Rates in your industry and region? 2018 turnover report. Available at: https://www2.salary.com/l/250572/2018-12-05/8slmkq (accessed 22 July 2021).

Sallam, M. and Winter, O. (2017) Egypt and Israel: forty years in the desert of cold peace. *Strategic Assessment* 20(3), 21–34.

Singh, R.P. and Kaur, H. (2016) Managing cross-cultural: issues and challenges in global organizations. *International Journal of Advance Research and Innovative Ideas in Education* 2(3), 4034–4038.

Song, S. (2020) Multiculturalism. Stanford Encyclopedia of Philosophy. Available at: https://plato.stanford.edu/multiculturalism/ (accessed 25 October 2022).

Spinks, N. and Moore, C. (2007) The changing workforce, workplace and nature of work: implications for health human resource management. *Nursing Leadership* 20(3), 26–41. DOI: 10.12927/cjnl.2007.19286.

Sütalan, Z. (2015) *Explaining Cold Peace in the Middle East: The Cases of Egyptian–Israeli and Jordanian–Israeli Peace Processes*. Doctoral dissertation, Graduate School of Social Sciences, Middle East Technical University, Ankara, Turkey. Available at: https://etd.lib.metu.edu.tr/upload/12618385/index.pdf (accessed 25 October 2022).

Swarnalatha, C. and Tephillah Vasantham, S. (2014) Cross-cultural management – an overview. *International Journal of Research in Management & Business Studies* 1(3), 9–10.

Tosti, D.T. (2002) Global fluency. In: Beaman, K. (ed.) *Boundaryless HR: Human Capital Management in the Global Economy*. Rector Duncan & Associates, Inc, Austin, Texas, pp. 109–119.

Zalzberg, O. (2020) The regional stakes of soured Israeli–Jordanian relations. *International Crisis Group*. Available at: https://www.crisisgroup.org/middle-east-north-africa/eastern-med terranean/jordan/regional-stakes-soured-israeli-jordanian-relations (accessed 27 October 2021).

9 Geotourism Development in the Middle East: A Comparative Study of Iran, Saudi Arabia, Oman and Jordan

Bahram Nekouie Sadry[1*], Kholoud Mohamed Abdel Maksoud[2] and Somayeh Zahabnazouri[3]

[1]University of Science and Culture, Tehran, Iran, Chairman of Geotourism Standing Committee, Ecotourism Scientific Association of Iran, Tabrīz, Iran; [2]Cairo University, Egypt; [3]Supporters of the Lut Desert NGO, Kerman, Iran

9.1 Introduction

In the past, nature-based tourism activities were primarily focused on living things (namely biotic nature attractions/biodiversity phenomena) (Sadry, 2009a). As a result, visiting natural areas and their unique flora and fauna gradually led to the development of ecotourism, wildlife tourism and other similar tourism types. In the 21st century, geotourism has emerged to deal with non-living parts of the natural environment. Geological features and landforms are examples of abiotic nature or geodiversity phenomena (Sadry, 2009a). As a result, geotourism today can be considered a new type of sustainable tourism that focuses on geosites and provides tourists with knowledge, environmental education and recreation (Santangelo and Valente, 2020).

A geosite/geological heritage site is a location with a specific geological or geomorphological value (aesthetic, uniqueness, educational, economic, etc.) (Reynard, 2009) and contains a distinctive form of abiotic natural feature. Some geoheritage sites, such as mining sites, are related to human activity and may be viewed as cultural geosites (Hose, 2006; Sadry, 2009a). A larger scientific community must investigate geoheritage to contribute to geotourism (Newsome and Dowling, 2018). Geotourism is a fundamental pillar in developing geoparks, which has resulted in regional investment, job opportunities and increased income generation among local residents (Farsani *et al.*, 2012; Sadry, 2021a).

A geopark is a united region that supports the sustainable conservation and use of geological heritage while promoting people's economic well-being (McKeever and Zouros, 2005). The 195 United Nations Educational, Cultural and Scientific Organization (UNESCO) Member States approved the creation of the UNESCO Global Geoparks designation on 17 November 2015 (UNESCO, 2015). This indicated the appreciation of world governments for the need for holistic conservation of magnificent geological sites and landscapes. Member States decided during UNESCO's 38th General Conference (IGGP) to support the legislation of a new worldwide programme,

*Corresponding author: Bahram.Sadry@usc.ac.ir

 Tourism Planning and Development in the Middle East (eds S. Kladou *et al.*)
DOI: 10.1079/9781789249149.0009

the International Geoscience and Geoparks Programme. This allows the organization to better represent today's socioeconomic issues in earth science and recognizes a prior network of geologically significant areas globally (UNESCO, 2015).

The Middle East has a large geodiversity that the region's governments have not addressed (Sadry, 2021a). This is due to limited geoheritage and geoconservation research, particularly the lack of studies identifying, inventorying and assigning value to this geodiversity (Errami *et al.*, 2015). This chapter mainly discusses the development of geotourism and the main concerns and implications of geoheritage and geotourism in the Middle East. It begins with an introduction and a review of the literature on geotourism, geoparks and other related topics, followed by a brief section outlining the case study methodology and clarifying some of the more important countries in the region that have not been included in this chapter. The next section is devoted to findings and discussion, focusing on Iran. It begins by describing the reasons that have made Iran a desirable place for developing geotourism through the introduction of the first UNESCO Global Geopark in the region and two other aspiring ones in Iran. Then, the discussion continues with coverage of Saudi Arabia, Oman and Jordan and the presentation of each case and its appropriateness for geotourism development and geopark establishment. Finally, managerial issues will be discussed, including implications, limitations and future research directions.

9.2 Review of the Literature on Geotourism and Geoparks

Geodiversity, geoheritage (represented by geosites and geomorphosites) and geotourism are all intrinsically linked. Geotourists interested in the Earth's geological history are interested in geosites and geomorphosites. Geomorphosites have a more complicated definition than geosites and are thought to have more geomorphological relevance. They are defined as sites that are particularly important for understanding the Earth's geomorphic processes scientifically and are of cultural, aesthetic and economic value (Brilha, 2018). The broader viewpoint underpinning the idea of geomorphosites allows for a more accurate assessment of the importance of geological and geomorphological sites to any geotourism project. Geomorphosites will help promote awareness of the importance of geomorphological and geoarchaeological sites and their features and socioeconomic significance (Reynard, 2009). Geodiversity must be viewed through a multidisciplinary approach that prioritizes human needs (and artefacts). The human aspect must also be considered when maintaining or developing archaeological sites within geosites. It can help scientists, students, visitors and other casual recreationalists by identifying geo-archaeological sites as protected sites.

A geosite can reach its potential as a tourism attraction if visits are adequately managed. Visitors are offered the chance to learn about geological features and processes during their visit. However, before developing a geosite for tourism purposes, two factors must be evaluated: (i) the geological value to humans of its geoheritage; and (ii) the geoconservation requirements, i.e. the requirements for its protection. Therefore, an evaluation must be conducted as part of the inventorying and cataloging of its artefacts in geotourism planning and management to determine a site's value (Štrba *et al.*, 2015; Brilha, 2018). Geosites should be viewed as deserving of preservation and protection, both against rapid natural degradation and destructive human activities, for the community and future generations (Schumann *et al.*, 2015).

Geotourism and geoparks have developed and been recognized as having significant potential to enhance long-term rural development and alleviate poverty globally (Piranha and Carneiro, 2009; Farsani *et al.*, 2012; Sadry, 2021a). A geopark aids in promoting culturally and environmentally sustainable socioeconomic growth and the improvement of people's living conditions and the rural environment. According to Farsani *et al.* (2012), one of the primary strategic goals of a geopark is to stimulate economic activity and long-term growth in the surrounding community. Geotourism also involves collaboration between the government, residents and the private sector, including local companies, outdoor enterprises tour agencies, restaurants and lodging. This partnership is financially sound and beneficial to all parties involved (Dowling, 2009).

9.2.1 A brief history of the geopark movement

Geoparks are a manifestation of the development of the geotourism industry in the 21st century. Briefly, according to Sadry (2020b), a geopark is a place full of geological heritage with a local community living nearby and which will contribute to tourism and geology education, while the principal livelihood of the local community is still derived from and based on geology. According to Jones (2008), the geopark concept was initially suggested at the Digne Convention in 1991 to safeguard and promote geological heritage (geosite/geoheritage site) and sustainable local development through a global network of territories involving geology of significant interest. Consequently, in 1997, the UNESCO Division of Earth Sciences developed the notion of a UNESCO Geoparks Program to support national and worldwide efforts in Earth heritage protection in response to the 'Declaration of the Rights of the Memory of the Earth'. Representatives from four European regions gathered in 2000 to discuss their economic development through geological heritage preservation and geotourism promotion. This meeting resulted in a convention declaring the foundation of the European Geoparks Network (EGN). The next essential step for the EGN was the signing of a formal collaboration agreement with UNESCO in 2001, which placed the Network under the organization's authority. So, under the patronage of UNESCO, the 17 existing European geoparks merged with eight new Chinese national geoparks to form the Global Network of National Geoparks in 2004 (Jones, 2008).

Gray (2004) predicted that by the year 2025 the number of geoparks would have risen to 500. As of 2022, 81 European geoparks (in 26 countries) and 169 Global Geoparks (in 44 countries) have been declared around the globe. In the face of this, Gray's (2004) prediction is wrong because the number of geoparks globally is half the figure he predicted for 2025. This circumstance calls for an investigation of the factors causing this 'inconsistency'. This study contributes to this investigation by exploring the Middle East region collectively. In the centre of the Middle East, Sadry (2012) suggested and highlighted the existence of at least 50 viable areas as potential Global Geoparks within Iran (Hajalilu and Sadry, 2012). Qeshm was the area of the first Iranian geopark and the first UNESCO Global Geopark in the Middle East, designated by UNESCO in 2005 by the late Dr Abdolazim Haghipour, former Director of the Iranian Geological Survey (Turner, 2008). At the Paris conference (21 March 2006), Qeshm Island in the Persian Gulf area of Iran was listed in the Global Geoparks Network (GGN).

9.2.2 UNESCO inscription and sustainability: geotourism and geoparks

Geotourism during the last 15 years has expanded from an unknown niche trend to a tourism planning and abiotic environmental protection approach, particularly with the UNESCO-sponsored Global Geopark movement in the 21st century (Sadry, 2021b). UNESCO Global Geoparks are a mechanism of international cooperation in which areas of the geological heritage of international significance support each other in engaging with local communities to promote awareness of that heritage and adopt a sustainable approach to the area's development through a bottom-up approach to conservation. These places can apply to UNESCO for designation as a 'UNESCO Global Geopark', drawing on the Organization's broader mission (UNESCO, 2022a).

In effect, 'in a philosophical manner, geotourism with the suffix (-ism) is an umbrella for geoparks ... but from the managerial viewpoint (especially for practitioners) geotourism is just one of the tourism activities (the main one) within geoparks among other "geopark tourism" activities' (Sadry, 2021b, p. 7)

A thorough understanding of geological heritage and tourism in the Middle East is crucial for a holistic strategy for long-term development. Geopark tourism might open up a wide range of economic development opportunities if it is included in long-term tourism plans.

With spectacular landscapes and stunning panoramas all around the Middle East, the region is dedicated to establishing itself as an unrivalled, world-class tourism destination. The Middle East's geoheritage provides a foundation for geotourism operations. Geo-conservation in the Middle East necessitates legislation and the development of geoparks allows Middle Eastern governments to do so under the auspices of UNESCO.

According to UNESCO (2022b), UNESCO Global Geoparks are subjected to a rigorous review of their operation and quality every 4 years to ensure their pristine state and sustainability. Two evaluators go on a field mission to re-validate the quality of the UNESCO Global Geopark. If the UNESCO Global Geopark continues to meet criteria based upon the field evaluation report, the area will be declared as a UNESCO Global Geopark for another 4 years (so-called 'green card'); if the criteria are no longer being met, the management body will be notified to take appropriate action within 2 years (so-called 'yellow card'). If it fails to meet the criteria within two years after receiving a 'yellow card', the area's designation as a UNESCO Global Geopark will be revoked (so-called 'red card'). The UNESCO Global Geopark under review must prepare the following as part of the revalidation process: (i) a one-page summary of the UNESCO Global Geopark to be submitted to the UNESCO Secretariat 1 year prior to the revalidation (by the end of July); and (ii) a progress report that used a template, self-evaluation and progress evaluation form to be submitted through the official channel (by the end of February at the latest).

9.2.3 The case studies: overview and methodology

With its different sects, the Muslim faith is the most widely practised religion in the Middle East, and Islam in the region's primary link between people. As a result, the Middle East has become known as a cradle of religious tourism and a hub of spiritual, Islamic and Halal tourism, all of which have experienced rapid growth in recent decades (Zamani Farahani *et al.*, 2019). The Middle East also has the potential to become a significant tourism destination for Muslims from all over the world due to its natural and cultural diversity. However, this potential has not yet been completely realized due to political, cultural and economic factors (Zamani Farahani *et al.*, 2019). In this study, the documentary research method (i.e. the use of extensive secondary data to support the opinions) helps analyse the backgrounds and initiatives of geoheritage in the Middle East. During the selection process of countries to be included in the documentary research method, Turkey and Egypt were excluded. The reason to exclude Egypt is the publisher's strategy, to consider a more geographical and less geopolitical approach, and North African cases will be considered in another volume. Turkey was eliminated because it is one of the European Union's leading partners and has ties to the EGN, although not officially part of the EU. The EGN's main objective is to ensure cooperation between geoparks for the protection of geological heritage and the promotion of sustainable development of their territories in Europe, including Turkey. Therefore, Saudi Arabia, Oman, Jordan and Iran are the focus of this study due to their background in recognizing geotourism, available data, and governmental or research accounts, but some other countries are mentioned on the map (Fig. 9.1) as they contain the most recent suggested Geopark areas (as a preliminary plan). They are based on: Sadry (2009b); Sabouri (2010); Fakhari (2014); Malkani (2014); Ahmadi (2015); Al Wosabi and Al-Aydrus (2015); Al-Ismaili (2019); Fu *et al.* (2021); Sadry *et al.* (2021); and UNESCO (2021).

Given the nature of geotourism, a brief overview of the climate and the geological setting is necessary. Countries of the region, politically referred to as the 'Middle East', span two continents, with varied and complex geology. Four case studies located in the Arabian Peninsula and the Eurasian Plate help exemplify this complex geology and potential for geotourism development. In this geological section, first, the most important tectonic events that helped form the geological features found in the regions included in these case studies are summarized. As can be seen from our case studies in Saudi Arabia, Oman and Jordan, the geology of the Middle East is worth knowing and studying. It also has benefits in geotourism and the

Fig. 9.1. Map shows the location of identified geoparks in the Middle East (excluding Turkey: see text). Source: Google Earth; Data SIO, NOAA, US Navy, NGA, GEBCO. Image: Landsat /Copernicus; annotations by chapter authors.

establishment of Global Geoparks (see Fig. 9.1). Some areas are newly being introduced such as Sabzevar Geopark in Razavi Khorasan Province or Bisotun Geopark (Sadry and Tavazo, 2022) in Kermanshah Province of Iran and in the studied countries that are not shown on the map.

9.3 Results and Discussion

9.3.1 Geoparks and geotourism development in Iran

Despite the existence of numerous geoheritage sites (geosites) in the region, there are only two officially registered geoparks in the Middle East region the UNESCO Global Geopark on Qeshm Island, which was designated in 2005 (Turner, 2008) and covers an area of 2063 km^2 and the newly designated Aras UNESCO Global Geopark. Another aspiring geopark in the region located in Iran is Tabas UNESCO Global Geopark, which suggests that the country is one of the few in the Middle East which is attempting to benefit from its geological diversity by developing geoparks. Geopark establishment gives us a great opportunity for sustainable tourism development; the development of other forms of tourism in geoparks are not only permitted but would add to the diversity of the geopark (Sadry, 2021a, p. 10). Iran has a unique geodiversity, as it is home to a wide range of geological phenomena. Nabavi (1999) spoke on Iran's geology and geotourism capabilities at the Geological Survey of Iran (GSI) annual symposium, stating that Iran is able to utilize and take advantage of such geoheritage (Amrikazemi, 2012). Following that presentation, in 2000, UNESCO issued a call for countries to recognize their geological heritage, prompting the GSI to to develop a geotourism project. This project resulted in the identification and introduction of Iran's unique geological phenomena from an aesthetic perspective, which was necessary to boost geotourism. Investigation and identification of Qeshm Island from a geotourism perspective, as well as creating a geopark registration proposal in Qeshm in 2002, are some of that programme's outcomes (Amrikazemi, 2012). Thanks to the GSI initiatives, Iran in 2012 was named among the countries that have placed a national priority on geological heritage and geoparks (Sadry, 2020b; Sadry *et al.*, 2021).

9.3.1.1 *Qeshm Island UNESCO Global Geopark*

On Qeshm Island, Palearctic and Oriental zoogeographical zones, as well as Afro-tropical, Oriental and Eurasian phytogeographical areas intersect, resulting in a great diversity of wildlife and vegetation. Various segments (e.g. high school students) of local communities are participating in geotourism activities in a great variety of ways and are deriving benefits from associating with this sustainable and diverse environment, which gives a genuine incentive for the community to protect geosites. Unfortunately, community involvement in the geological project only seriously started about 10 years after establishing the Qeshm Island Geopark. The Qeshm Island Geopark has also developed a special teaching programme based on the framework of school science textbooks. It motivates children to learn about the Earth, instills a sense of responsibility in them, for the planet's resources and encourages working in groups.

Qeshm Island Geopark, Iran's main geopark, was delisted from the GGN in 2012 in spite of the developments that had taken place in the infrastructure and in pursuit of the goals and functions of a geopark. It is assumed that the weakness was rooted in the culture of the region, which did not allow the development of western style geotourism; the poor participation of the geopark in GGN has been identified as a secondary reason (Turner, 2018, personal communication, cited in Sadry, 2021c). However, Qeshm Island was listed once again in 2017 and received the opportunity to demonstrate its potential in the GGN. However, being back on the list does not mean that the problems are resolved. In fact, according to a research study by Sadry (2020a), just 10% of travellers arriving in Qeshm have visited the island's geosites in the last 5 years. The rest came to the island for shopping (2015–2019), this demonstrates the lack of geopark visibility at the national level. It also demonstrates the lack of proper interpretation facilities, geotrails and geo-activities, etc. (Sadry, 2020a). The results of Sadry's study indicated that the current geotourism management system of Qeshm Island geopark benefited very little from geological interpretation provision for visitors. Fortunately, Qeshm Island Geopark was confirmed and awarded the Green Card by UNESCO evaluators in a re-validation process in 2022 (for the second time after being delisted in 2012). This can be a sign of glimmers of hope in development of geotourism in the Middle East.

9.3.1.2 *Aras UNESCO Global Geopark*

Aras UNESCO Global Geopark (Aras UGGp) includes a number of geosites. The most important geological feature of global significance in this geopark is the Permian–Triassic Boundary, the stratigraphic position of the greatest extinction event known in Earth's history. This site offers geoscientists the opportunity to investigate uninterrupted, highly fossiliferous successions in carbonate to claystone-dominated facies, which document this event, as well as the crucial connection of this site to other well-studied Permian–Triassic boundary stratotype sites around the globe. From a cultural viewpoint, Aras hosts Cultural World Heritage Sites, remnants of ancient civilizations, and historical buildings and locations connected to the Second World War. Aras hosts several local events during the year. These events familiarize tourists with the culture and art of local peoples living inside the geopark territory. The fauna and flora are also unique and, in some cases, rare. Three protected areas inside the geopark provide a virgin and relaxing environment for those who are looking for an opportunity to escape the city and have a restful and joyous communion with nature. Thus, Aras offers geological, natural and cultural attractions (see Fig. 9.2 for the territory coordination).

In 2011, the GGN denied the designation of this geopark based upon 10 arguments, the most important include: (i) unclear geographical boundaries; (ii) lack of coordinated and systematic management; (iii) under-developed geosites; (iv) lack of interpretational facilities; (v) lack of visibility; (vi) lack of engagement of local people in geopark activities; and (vii) the lack of any action plan after 7 years of trial and error. Recently, the UNESCO Global Geoparks Council chair offered a general action plan for Aras Geopark, which was then used as the primary strategy of the geopark in attaining its future goals (e.g. achieving sustainability throughout interpretation provisions). According to

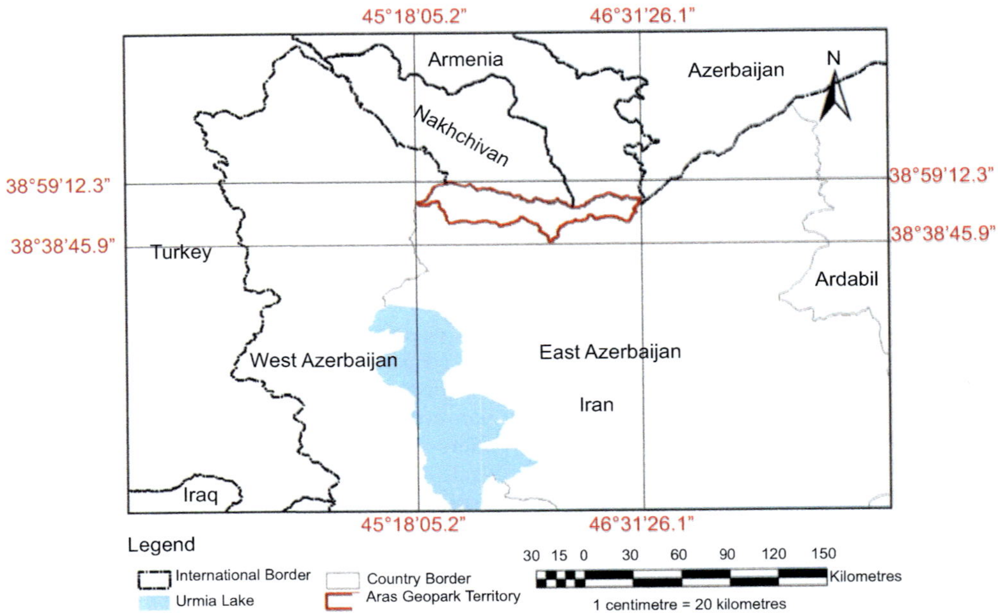

Fig. 9.2. The 2019 territory coordination of Aras Aspiring UNESCO Global Geopark (aUGGp). Source: Application Dossier of Aras Aspiring UNESCO Global Geopark (2019). Figure used with permission of Dr Fatemeh Moniri Alamdari.

this action plan, a number of changes were implemented. As a result, in 2019, a letter of interest to UNESCO was issued to proclaim Aras Geopark as a new applicant for participation in the Global Network (Dossier of Aras Aspiring UNESCO Global Geopark, 2019). Fortunately, Aras Geopark was awarded the Green Card by UNESCO evaluators in 2022 and a new glimmer of hope appeared on the horizon of geotourism development in the Middle East.

9.3.1.3 Tabas Aspiring UNESCO Global Geopark

Tabas Geopark is located in eastern Iran, in the northwestern portion of South Khorasan Province (Fig. 9.3). In 2017, the Tabas Geopark Project was proposed to the local government after the GSI conducted preliminary geological studies and ensured the presence of geotourism potential in this region of Iran. This geopark has unique geological features and offers various geological attractions. The most complete Paleozoic sediments found in Iran and West Asia and more than 20 types and reference sections of various formations have motivated experts to propose the county as a geological paradise and fossil museum of Iran. Hence, it is possible to see and study all the geological attractions, including tectonics, sedimentology, sedimentary petrology, paleontology, economic geology, petrology, various mineral resources etc. in the region. The rapid industrial and economic growth in Tabas has created a motivation for the protection of geological heritage and other valuable sites. Revival of the original culture and customs of the region and their promotion in the local community are very important. Therefore, according to its goals and philosophy, the Tabas Geopark could act as a barrier against the destruction or disturbance of land use of this geological heritage.

Education and conservation programmes dealing with the region's heritage are at the top of the geopark's activities. The formation of the National Network of Geoparks of Iran, a memorandum of cooperation with various universities and scientific associations, the construction of a Geoheritage museum, standardization of

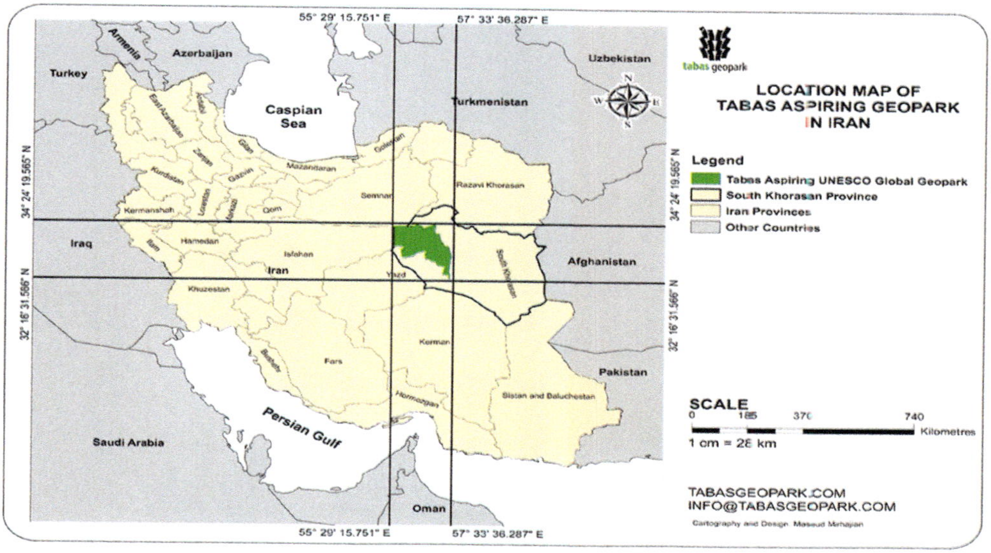

Fig. 9.3. The territory coordination of Tabas Aspiring UNESCO Global Geopark. Source: Application Dossier of Tabas Aspiring UNESCO Global Geopark (2021). Figure used with permission of Dr Vesal Yahya-Sheibani.

geosites and the compilation of the Geoheritage Atlas of Tabas are some of the activities being conducted. Obviously, the limited knowledge of geologists in the field of tourism and the lack of experts, especially in the field of geoheritage interpretation, is a key problem for geotourism site development in Tabas. In 2017, the chair of the UNESCO Global Geoparks Council proposed a general action plan for Tabas Geopark, which housed the principal geopark strategy to approach its future goals. The geopark idea was then studied and endorsed by local officials. During the same year, the board of trustees was formed, consisting of the local governor, the CEO of Tabas Parvadeh Coal Company and other local trustees.

The board assigned the task of carrying out the Tabas Geopark Project to the non-governmental/non-profit institute of 'Tabas Geopark'. With further development, that institute joined Tabas Parvadeh Coal Company as a new department. The application dossier of Tabas Aspiring UNESCO Global Geopark was sent to UNESCO for global registration in November 2021, and according to the studies, 39 geosites have been introduced in this geopark (including cultural, geo-archaeological and secondary geosites (Fig. 9.4). Tabas Geopark currently awaits visits from UNESCO evaluators and confirmation (green card for 4 years) or a conditional approval (yellow card to rebuild and compensate in 2 years' time).

9.3.2 Saudi Arabia

The region of the Kingdom of Saudi Arabia encompasses a considerable portion of the Arabian Peninsula. The potential geological legacy of several cities and locations in the kingdom has been investigated. Buraydah' (the capital of Al-Qassim Region in north-central Saudi Arabia) features all the necessary infrastructure needed to support geotourism. Due to its various tourist attractions and infrastructure, it scores well on the 'valorization' criterion for geotourism (Doktor *et al.*, 2015). Qusiaba is being actively developed for geotourism by the Saudi Commission for Tourism and National Heritage (Saudi, 2018).

The Al-Qassim region of Saudi Arabia, according to Kaminski *et al.* (2019), shows a lot of promise for geotourism, with a range of

Fig. 9.4. View of Nayband cultural geosite and Nayband mountains, south of the Tabas Aspiring Geopark. Photograph used with permission of Ehsan Zamaniyan.

remarkable geological localities that are well exposed and accessible. Paleozoic strata are also among the most significant geological units in terms of the Middle East's natural resources. Because they include rare fossils and microfossils that provide new insights into the development of land plants and marine microorganisms, they have tremendous educational value for geoscience students and professionals. In addition, the stratigraphic strata identified in the Al-Qassim area may help geoscientists better comprehend the history of the Gondwana continental shelf, notably the most recent Ordovician glacial episode (a geological term for an ice age interval). The 'Ridges' are regarded as part of the Asir Mountains' geological and geomorphological history (Elassal, 2020), and because of the prevalence of granitic lithology, alpine morphology and weathering-related formations, these ridges have substantial geotouristic potential. For generations, geotourists have been drawn to the pyramidal peaks (Al Soudah, Al Far'aa, Musharaf, Tahau, Riman, Munea, Athrib, Al-Sayyrah and Qarn Mijal tops) and steep slopes that contrast sharply with the lower limitrophe units. Unique formations add to the geomorphological landscape.

The HAMVG (Harrat Al-Madinah Volcanic Geopark) should be established as the Kingdom of Saudi Arabia's first geopark, according to Moufti and Németh (2013). The success of this project will likely have an impact on future geoconservation and geo-educational efforts in the Kingdom, and it might serve as a model for similar measures elsewhere in the Arabian Peninsula. The proposed geopark is predicted to have a substantial influence on geotourism. As a result of the interplay between harsh climate and volcanic landforms associated with intra-continental monogenetic volcanism, Harrat Al-Madinah's volcanic geology boasts a diverse range of volcanic features and an impressive volcanic landscape, and natural wonders of geo-educational and geotouristic value. Briefly, the Al-Medina volcanic field (with at least nine Cenozoic volcanic fields in western Saudi Arabia) has considerable potential as an aspiring geopark in the future (Fu *et al.*, 2021) (see Fig. 9.1).

9.3.3 Oman

The Sultanate of Oman is located in the Arabian Peninsula's northeastern corner. Large portions of old Tethyan oceanic crust were obducted onto the Arabian plate during the Late Cretaceous era, resulting in one of the world's best-exposed and studied ophiolites (Searle and Cox, 1999). Perhaps the best-known example of Oman's geological history is the Samail Ophiolite. On the other hand, Oman has many more interesting or unusual geological characteristics, including the Neoproterozoic glacial deposits (Allen, 2007); surface penetrating salt domes (Al Siyabi and Newall, 2005); large sand deserts (Goudie *et al.*, 1987); a 3000-metre-high mountain range with outstanding gorges like Wadi Nakhr (Kusky *et al.*, 2005); and the outcropping Moho (Kusky *et al.*, 2005) are some examples (Boudier and Nicolas, 1995). Due to the country's current climate, which spans from dry to semiarid zones, soil development and plant cover are severely limited. During the Pleistocene, the region was characterized by significantly higher rainfall and much more lush vegetation. As a result, geology is becoming more accessible, attracting an increasing number of geoscience students to study geology in the field, as well as visitors who come to appreciate the beautiful environment. In terms of environmental preservation in general, however, Oman has only recently begun to acknowledge that landscape degradation is spreading and posing a threat to biodiversity. The majority of the country's economic progress has occurred in the last 40 years. Large infrastructure projects are being built to fulfil the needs of continuing land and resource development. As a result, the geological heritage of Oman is now in danger of being lost due to a lack of information. In response to this, the Third Global Geotourism Conference was held in Oman from 30 October to 1 November 2011, attracting up to 300 delegates, including tourism experts and international agencies.

The Ministry of Tourism invited the Geological Society of Oman (GSO) to conduct a feasibility study for the sustainable development of geoparks in the Sultanate of Oman in 2019 (Al-Ismaili, 2019). According to Al-Ismaili (2019), the study was completed in September 2019 and evaluated and ranked seven potential geoparks. It concluded that Oman has great potential for geotourism and recommended two areas for immediate development as geoparks, with the possibility of adding more areas in the future. The government is very interested in developments that preserve Oman's unique identity while also enabling and adding value to local communities. So, the Al Huqf Area in Central Oman and Samail Ophiolite (the Oman ophiolite, known as Samail Ophiolites, are the most stunning exposed oceanic lithosphere in the world), and are intriguing examples of Oman's geological areas and hence potential geoparks (Fig. 9.1). Yet practical progress for the establishment of the geoparks and working on geological conservation is very slow.

9.3.4 Jordan

Jordan's geological and geomorphic features are diverse and rich (Allan, 2012). Its territory includes the lowest point on the planet, at 400 m below sea level (JICA and MOTA, 1996). It also has great biodiversity, and most geosites can be linked to historical, archaeological and cultural locations, and even sites sacred to Muslims, Christians and other religions, e.g. Petra, the Nabatean capital and the Byzantine Christian site of Madaba, and numerous archaeological sites in Wadi Feynan, including Yarmoukian sites such as Wadi Ghaweir.

In 2014, Jordan worked to propose its first geopark in a 66 km^2 area of Wadi Al Mujib, an area of rich geological diversity, which exposes the geological record in the area south of the Dead Sea, from Precambrian coastal marine sandstones to Quaternary terrestrial deposits. The proposed park meets the scientific requirements in terms of rarity, outstanding scientific value and beauty. In general, however, the geoheritage and geosites in Jordan have confronted many barriers, including accessibility and infrastructure issues, limited information and awareness of the local geoheritage and its value, issues of interpretation and problems of geoheritage sustainability (Dowling and Allan, 2018). In addition to these, Jordan, like other nations (e.g. Iran), requires a large number of supportive people to nurture the geopark movement in the country, and thus far, these are not available.

9.4 Conclusions

This study demonstrates that a detailed understanding of geoheritage is critical for a comprehensive plan for sustainable tourism-related development. Geoheritage sites frequently have a complex 'story to tell', which spans hundreds of millions of years. As a result, geosites are extremely useful in both education and science (Schumann *et al.*, 2015). The value of a geological feature can have a considerable impact on its management, protection and educational elements, as well as upon the effective long-term development of the area in which it is located. The concept of geoheritage, or the aspects of the Earth that people value, must be widely adopted and places designated as having such value should be protected. When geological characteristics are recognized as potential development geosites, geoconservation, or the sensible use of such sites, must take precedence. These areas can only be exploited for geotourism when the necessary and suitable protective measures have been implemented. To guarantee that the sites' (geological features) advantages (environmental, social and economic) outweigh any negative consequences, optimal and sustainable geotourism necessitates detailed planning and appropriate site access control.

Because geology's value for tourists does not constitute a formal nomination for historic designation, geoheritage evaluation is merely one step in the process of recognition and preservation. According to White and Wakelinking (2014), a nomination is contingent on individuals or groups wishing to participate in the official process of bringing a site forward for appreciation, promotion and recognition of geoheritage. As a consequence, geotourism can assist in promoting public awareness of geoheritage as long as we don't lose sight of the focus, which is to enjoy but not neglect their adequate preservation.

Geoparks are essential for any given region's cultural and natural heritage, particularly in the Middle East. The geotourism industry is a relatively new idea with much promise not only in the Middle East but also globally, especially when a single location includes many geosites of varying scales, as in our case studies.

In order to achieve sustainable geotourism development, an approach is provided in the transition from geomorphosite evaluation to geotourism interpretation to (i) maintaining the value of geoheritage and (ii) coping with the rising dangers connected with human activities (Neches, 2013). Initially, the most essential part of this approach was understood as geology and geomorphology (Hose, 2008), but ideas have now evolved to include the more significant subject of 'landscape interpretation' (Newsome and Dowling, 2018). Local communities require non-formal education initiatives in addition to tourist education (interpretation) activities (Sadry *et al.*, 2022). In geotourism zones, the importance of community participation (Boley *et al.*, 2014) is connected to sustainable development and poverty eradication (Reimold *et al.*, 2006). The geopark area establishes a working setting in which nature and humans cohabit with the aim of geoconservation in the context of geotourism (Yolal, 2012).

The environment, geology and nature of the Middle East region have unique features. With the development of geotourism, the geoparks will provide a great opportunity to introduce and promote the geology and scenic landscape of this region, as well as to provide a sustainable source of income for local peoples in the Middle East. Geotourism is a subsector of sustainable tourism that has not yet been managed correctly in countries where their biodiversity is limited due to climatological components and which are part of the dry or semi-dry areas that comprise the less-developed countries of the world (Sadry, 2021c).

Geotourism development due to the existence of excellent geological and geomorphological outcrops is an alternative to agricultural and industrial expansion in regions with arid and semi-arid conditions and water-scarce environments of the Middle East (Suppan *et al.*, 2008, p. 47). Due to the scarcity of water, agriculture usually relies upon a costly irrigation infrastructure (Hasanean, n.d.). Because of its arid conditions and restricted water supply, the Middle East is particularly sensitive to climate change. Instead of developing water-based enterprises, government-scale geopark development might be an effective way to alleviate poverty in the Middle East.

9.5 Managerial Implications

Discovering innovative ways to improve resource management is just one example of how improving resource efficiency can enhance the economic potential of this region. The Middle East is the best region to promote geotourism because of its arid and semi-arid climatic conditions, visible outcrops and amazing geo-phenomena. The Middle East offers stunning landscapes and cultural landmarks, as well as a world-renowned geological history of rocks, strata, minerals, fossils and other unique geoheritage sites (Sadry, 2012). As a result, some tourism-related recommendations suggest a Middle East Geoparks Network (MGN) should be established with the purpose of exploring, inventorying and valorizing the region's inherent rich geodiversity. The MGN should create awareness among the local community and decision makers about the necessity of sustainable use and management of geoheritage for the benefit of local socio-economic sustainable development aims by promoting both geotourism and the formation of geoparks. Geotourism development is a multidisciplinary subject; despite their skills, a geologist or geomorphologist would be unable to act alone on the basis of research. Teamwork appears to be crucial to understanding and promoting geotourism among professionals from various disciplines in transdisciplinary domains (and even those who have been trained in various disciplines) (Sadry *et al.*, 2021).

As a result, a one-dimensional approach to such activities would result in a destructive managerial perspective, sometimes known as 'toxic management'. It indicates that teamwork among professionals from diverse disciplines and fields is crucial to comprehending and promoting geotourism. Heritage interpretation, particularly geoheritage interpretation, is both a science and an art that is less widely known in developing countries. Their tourist development is unfamiliar with such facilities. At the national level, heritage interpretation and science generalization are essential concerns that both academics and executive directors are less familiar with. As a consequence, one of the barriers to the development of geoparks in developing countries is a lack of true recognition that, first and foremost, this type of tourism is knowledge-based and that the monopolization of the sector's activities is tied to public sector budgets. As a result, when governmental authorities' affiliates refuse, geoparks fail to allow the presence and participation of other knowledgeable people, including specialists (Sadry, 2021a). Furthermore, not just a small group of people, but many good and motivated individuals are needed in the Middle East geopark movement. The obstacles in Iran primarily involve the managerial and consultant and observation and decision-making team from GSI. For instance, the rejection of Aras Geopark in 2011 was due to the weak consultations and managerial implications which prevented the case from proceeding successfully, similar to the case of Qeshm Island before its de-listing from GGN in 2012.

Raising awareness in all relevant institutions, increasing national implementation through the integration of local and regional stakeholders, and establishing geoparks are all critical to improving geoheritage science and establishing geoparks, which are among the geoheritage and geoconservation challenges facing the Middle East in the coming decades. Sadry (2021c) believes that geotourism and geoparks have not yet been acknowledged appropriately in impoverished and far-off regions among managers and policy makers. Due to a lack of legislation allowing for the creation of geosite inventories, as well as the conservation and economic use of Middle Eastern geoheritage, the creation of geoparks would provide an opportunity to create a local inventory and local laws to protect the geoheritage in various Middle Eastern countries. UNESCO has developed a geopark brand and distributes labels to the network's partners. Combining the geopark brand with the Halal brand (Halal tourism services) provides a fantastic tourism promotion opportunity, at least in the Middle East region.

To conclude, the development of a geopark can contribute to the development of tourism in the region and, as a result, benefit the economy. In 2017, Qeshm, the first geopark in the Middle East, was relisted as a Global Geopark, giving it a second chance to thrive and, in 2022, the park joined the global network. The exceptional quality of these evaluation missions, as well as the evaluators who conduct them, will ensure the quality of this new and prestigious UNESCO title for the next 4 years. Also, the aspiring UNESCO Global Geopark in Iran (Tabas) is awaiting visits by two UNESCO evaluators. However, Covid-19 delayed the evaluators' visits and assessment schedules. Still, it is worth mentioning that while preparing this book chapter, following the visit of UNESCO evaluators in 2022, Aras aspiring geopark joined the UNESCO Global Geopark network.

9.6 Future Research Directions

Some of the most important future directions of this study are: (i) on a worldwide and national basis, build and expand communication with tourism agencies and practitioners, and seek out new geotourism research, management and marketing opportunities, particularly as part of long-term tourism projects; (ii) improve stakeholders' 'geological' understanding, notably in the case of potential geotourism destinations; (iii) adopt modern technologies for improved geological site interpretation, increased education and awareness, and less environmental effect is also a priority; (iv) encourage conservation, interpretation and protection of geological characteristics.

The directives indicated above were accepted in November 2011 at the Sultanate of Oman Global Geotourism Conference and are applicable to all Middle Eastern countries.

Acknowledgements

We are thankful to Dr Vesal Yahya-Sheibani, Director of Tabas Geopark, and her respected colleague, Dr Musavian (Chairman of the Board of Directors of Tabas Aspiring UNESCO Global Geopark) and also Dr Fatemeh Moniri Alamdari, Director of Aras Aspiring UNESCO Global Geopark, for providing us with some information and pictures to be used in this chapter. Also, we are thankful to Zahra Tavazo for drawing the map used in this chapter.

We are also very grateful to Prof. Peter E. Wigand and Fatemeh Fehrest for their help in improving this chapter.

References

Ahmadi, A. (2015) Evaluation and feasibility study of geomorphosites to preserve geomorphological heritage: development of geomorphotourism and registration of national-global geopark (area of study: Uramanat [Hawraman] region). PhD dissertation, University of Isfahan, Isfahan, Iran.

Al-Ismaili, I. (2019) Oman geoparks and geotourism, geological society of oman publication. *Al-Hajar Magazine* 14–24. Available at: http://www.gso-oman.org/wp-content/uploads/2017/12/Al-Hajar-Magazine-26-edition.pdf (accessed 25 October 2022).

Allan, M. (2012) *Geotourism: Toward A Better Understanding of Motivations for A Geotourism Experience – a Self-Determination Theory Perspective*. LAP LAMBERT Academic Publishing, Saarbrücken, Germany.

Allen, P.A. (2007) The Huqf Supergroup of Oman: Basin development and context for Neoproterozoic glaciation. *Earth-Science Reviews* 84, 3–4. DOI: 10.1016/j.earscirev.2007.06.005.

Al Siyabi, H. and Newall, M.J. (2005) The salt domes of the interior northern Oman [unpublished paper].

Al Wosabi, M. and Al-Aydrus, A.A. (2015) Dinosaur footprint sites in Arhab area: an aspiring geopark in Yemen. In: Errami, E., Brocx, M. and Semeniuk, V. (eds) *From Geoheritage to Geoparks: Geoheritage, Geoparks and Geotourism*. Springer, Cham, Switzerland, pp. 167–182.

Amrikazemi, A. (2012) Geoheritage studies in Iran. In: *Geoscience and Mining Monthly Magazine* (Geoheritage Special Edition: Geodiversity-Geoconservation-Geopark-Geotourism). Geological Survey and Mineral Exploration of Iran (GSI), pp. 5–7.

Boley, B.B., McGehee, N.G., Perdue, R.R. and Long, P. (2014) Empowerment and resident attitudes toward tourism: strengthening the theoretical foundation through a Weberian lens. *Annals of Tourism Research* 49, 33–50. DOI: 10.1016/j.annals.2014.08.005.

Boudier, F. and Nicolas, A. (1995) Nature of the Moho transition zone in the Oman ophiolite. *Journal of Petrology* 36(3), 777–796. DOI: 10.1093/petrology/36.3.777.

Brilha, J. (2018) Geoheritage: inventories and evaluation. In: Reynard, E. and Brilha, J. (eds) *Geoheritage: Assessment, Protection, and Management*. Elsevier, Amsterdam, pp. 69–86.

Doktor, M., Miśkiewicz, K., Welc, E.M. and Mayer, W. (2015) Criteria of geotourism valorization specified for various recipients. *Geotourism* (42–43), 25–38. DOI: 10.7494/geotour.2015.42-43.25.

Dowling, R.K. (2009) New challenges with geotourism. In: Netode Carvalho, C. and Rodrigues, J. (eds) *Proceedings of the VIII European Geoparks Conference Idanha-a-Nova*. 4–6 September, 2009.

Dowling, R.K. and Allan, M. (2018) Who are geotourists? A case study from Jordan. In: Dowling, R. and Newsome, D. (eds) *Handbook of Geotourism*. Edward Elgar Publishing, Cheltenham, UK, pp. 76–86.

Elassal, M. (2020) Geomorphological heritage attractions proposed for geotourism in Asir mountains, Saudi Arabia. *Geoheritage* 12(4), 1–18. DOI: 10.1007/s12371-020-00505-z.

Errami, E., Gabi, S., Nasser, E., Hasina, N.R., Abderrahmane, B. *et al*. (2015) Geoheritage and geoparks in Africa and the Middle-East: challenges and perspectives. In: Errami, E., Brocx, M. and Semeniuk, V. (eds) *From Geoheritage to Geoparks: Case Studies from Africa and Beyond*. Springer, Cham, Switzerland, pp. 3–23. DOI: 10.1007/978-3-319-10708-0_1.

Fakhari, S. (2014) Geomorpholgy modeling of geoparks for sustainable development (case study: damavand). PhD thesis, Kharazmi University, Tehran, Iran.

Farsani, N.T., Coelho, C., Costa, C. and Neto de Carvalho, C. (eds) (2012) *Geoparks and Geotourism: New Approaches to Sustainability for the 21st Century*. BrownWalkerPress, Boca Raton, Florida.

Fu, H., Fu, B., Shi, P. and Zheng, Y. (2021) International geological significance of the potential Al-Medina volcanic UNESCO global geopark project in Saudi Arabia revealed from multi-satellite remote sensing data. *Journal of Heritage Science* 9(1), 149. DOI: 10.1186/s40494-021-00610-4.

Goudie, A.S., Warren, A., Jones, D.K.C. and Cooke, R.U. (1987) The character and possible origins of the aeolian sediments of the Wahiba Sand Sea, Oman. *The Geographical Journal* 153(2), 231. DOI: 10.2307/634875.

Gray, J.M. (2004) *Geodiversity: Valuing and Conserving Abiotic Nature*. John Wiley & Sons Ltd, Chichester, UK.

Hajalilu, B. and Sadry, B.N. (2012) *Geotourism*. The Payame Noor University Press, Tehran, Iran [in Persian].

Hasanean, H.M. (n.d.) Middle east meteorology. In: *Encyclopedia of Life Support Systems (EOLSS); UNESCO-EOLSS*. Developed under the auspices of the UNESCO, Eolss Publishers, Paris, France. Available at: http://iprc.soest.hawaii.edu/users/yqwang/Middle-east-meteorology.doc (accessed 25 October 2022).

Hose, T.A. (2006) Geotourism and interpretation. In: Dowling, R. and Newsome, D. (eds) *Geotourism*. Elsevier Butterworth-Heinemann, Oxford, UK and Burlington, Massachusetts, pp. 221–241.

Hose, T.A. (2008) Towards a history of geotourism: definitions, antecedents and the future. In: *The History of Geoconservation*. Geological Society, London, Special Publications 300, pp. 37–60. DOI: 10.1144/SP300.5.

JICA and MOTA (1996) *The Study on the Development Plan in the Hashemite Kingdom of Jordan*. JICA and MOTA, Amman.

Jones, C. (2008) History of geoparks. In: Burek, C.V. and Prosser, C.D. (eds) *The History of Geoconservation*. Geological Society, London, Special Publications 300, pp. 273–277. DOI: 10.1144/SP300.21.

Kaminski, M.A., Garrison, T.F. and Yoder, C.E. (2019) The paleozoic formations of the Al Qassim province in Saudi Arabia as potential sites for geotourism. *Geotourism/Geoturystyka* 58–59(1), 1. DOI: 10.7494/geotour.2019.58-59.1.

Kusky, T., Robinson, C. and El-Baz, F. (2005) Tertiary–quarternary faulting and uplift in the northern Oman Hajar mountains. *Journal of the Geological Society* 162(5), 871–888. DOI: 10.1144/0016-764904-122.

Malkani, M.S. (2014) Terrestrial ecosystem from the Mesozoic Geopark of Pakistan. In: *Proceedings of 6th International UNESCO Conference on Global Geoparks*, Saint John, New Brunswick, Canada, 19–22 September.

McKeever, P.J. and Zouros, N. (2005) Geoparks: celebrating earth heritage, sustaining local communities. *Episodes* 28(4), 274–278. DOI: 10.18814/epiiugs/2005/v28i4/006.

Moufti, M.R. and Németh, K. (2013) The establishment of a volcanic geopark in the kingdom of Saudi Arabia: Harrat Al Madinah Volcanic Geoprk. In: Buchner, J., Rapprich, V. and Tietz, O. (eds), *Basalt 2013–Senckenberg Scientific Conference*, Goerlitz, Germany, Abstract & Excursion Guides, April 24–28, 2013, pp. 203–204.

Nabavi, M.H. (1999) *Geotourism (Geological-Tourism): The Proceedings of the 18th Symposium on Earth Sciences*. Geological Survey and Mineral Exploration of Iran, pp. 1039–1049.

Neches, I.M. (2013) From geomorphosite evaluation to feotourism interpretation. Case study: the sphinx of Romania's southern Carpathians. *Geojournal of Tourism and Geosites* Year VI 12(2), 160.

Newsome, D. and Dowling, R. (2018) Geotourism and geoheritage. In: Reynard, E. and Brilha, J. (eds) *Geoheritage: Assessment, Protection, and Management*. Elsevier, Amsterdam, The Netherlands, pp. 305–321.

Piranha, J.M. and Carneiro, C.D.R. (2009). O ensino de geologia como instrumento formador de uma cultura de sustentabilidade. *Brazilian Journal of Geology* 39 (1), 129–137.

Reimold, W.U., Whitfield, G. and Wallmach, T.H. (2006) Geotourism potential of southern Africa. In: Dowling, R. and Newsome, D. (eds) *Geotourism*. Butterworth-Heinemann, Oxford, UK, pp. 42–62.

Reynard, E. (2009) The assessment of geomorphosites. In: Reynard, E., Coratza, P. and Regolini-Bissig, G. (eds) *Geomorphosites: Assessment, Mapping and Management*. PfeilVerlag, Munich, Germany, pp. 63–72.

Sabouri, T. (2010) Investigation and studies of geomorphology in land management of Gilan in order to create Dorfak and Deylamangeoparks, providing sustainable geotourism solutions. PhD dissertation, The Islamic Azad University, Science and Research Branch, Tehran, Iran.

Sadry, B.N. (2009a) *Fundamentals of Geotourism: With Special Emphasis on Iran*. SAMT Publications, Tehran, Iran.

Sadry, B.N. (2009b) Nomonation of suitable area for geopark and geotourism development in Jolfa and Varzeghan regions in Iran. In: Brilha, J., Pereira, D. and Pereira, P. (eds) *Intensive Course on Geodiversity and Geological Heritage Assessment Workbook*. University of Minho, Minho, Portugal.

Sadry, B.N. (2012) [Halal] Geopark brand in the middle east: a potential lever for socio-economic growth. In: *Proceedings of the First Geological Conference of Kurdistan, GEOKURDISTAN-2012*, University of Sulaimani, Sulaimani, Kurdistan Region, Iraq, 14–16 November, pp. 31–32.

Sadry, B.N. (2020a) Developing an educational model for geotourism management in geoparks; the case: Qeshm Island UNESCO global geopark. PhD dissertation, The Islamic Azad University, Science and Research Branch, Tehran, Iran.

Sadry, B.N. (2020b) *An Introduction to Geological Heritage of Iran: Geosites and Geomorphosites*. Co-publications by Mahkame Publishers and Iranian Association of Geomorphology, Tehran, Iran [in Persian].

Sadry, B.N. (2021a) The future of geotourism. In: Sadry, B.N. (ed.) *The Geotourism Industry in the 21st Century: the Origin, Principles, and Futuristic Approach*. Apple Academic Press, Palm Bay, Florida, pp. 535–552. DOI: 10.1201/9780429292798.

Sadry, B.N. (2021b) The scope and nature of geotourism in the 21st century. In: Sadry, B.N. (ed.) *The Geotourism Industry in the 21st Century: The Origin, Principles, and Futuristic Approach*. Apple Academic Press, Palm Bay, Florida, pp. 3–21. DOI: 10.1201/9780429292798.

Sadry, B.N. (2021c) *The Geotourism Industry in the 21st Century: the Origin, Principles, and Futuristic Approach*. Apple Academic Press, Palm Bay, Florida.

Sadry, B.N. and Tavazo, Z. (2022) Introducing and assessing the tectonic geoheritage in the sahne-harsin region of kermanshah on the west of iran. *Tectonics Journal* 5(17), 92–117 (in Persian).

Sadry, B.N., Ghourchian, N.G. and Mohammad-Davoudi, A.H. (2022) Developing a conceptual model for heritage interpretation in the geoparks based on authentic methods in the world. *Journal of Tourism Planning and Development* 30, 69–85.

Sadry, B.N., Mohammadi-Aragh, A., Fehrest, F., Bayatani, A. and Haji-Moradi, A. (2021) Identifying and assessing geodiversities around Takht-e Soleyman world heritage site to propose the territory as the Third Geopark in Iran. In: Singh, R., Wei, D. and Anand, S. (eds) *Global Geographical Heritage, Geoparks and Geotourism: Geoconservation and Development*. Springer, Singapore, pp. 15–42.

Santangelo, N. and Valente, E. (2020) Geoheritage and geotourism resources. *Resources* 9(7), 80. DOI: 10.3390/resources9070080.

Saudi, G. (2018) Qusaiba: a window to Saudi geography. Available at: https:// saudigazette.com.sa/article/527807 (accessed 30 July 2020).

Schumann, A., Muwanga, A., Lehto, T., Staudt, M., Schlüter, T. *et al*. (2015) Ugandan geosites. *Geology Today* 31(2), 59–67. DOI: 10.1111/gto.12089.

Searle, M. and Cox, J. (1999) Tectonic setting: origin, and obduction of the Oman ophiolite. *Geological Society of America Bulletin* 111(1), 0104. DOI: 10.1130/0016-7606(1999)111<0104:TSOAOO>2.3.CO;2.

Štrba, L., Rybár, P., Baláž, B., Molokáč, M., Hvizdák, L. *et al*. (2015) Geosite assessments: comparison of methods and results. *Current Issues in Tourism* 18(5), 496–510. DOI: 10.1080/13683500.2014.882885.

Suppan, P., Kunstmann, H., Heckl, A. and Rimmer, A. (2008) Impact of climate change on water availability in the Near East. In: Zereini, F. and Hötzl, H. (eds) *Climatic Changes and Water Resources in the Middle East and North Africa*. Springer-Verlag, Berlin and Heidelberg, Germany.

Turner, S. (2008) Geotourism': where wise birds stay awhile. *GeoExpro Magazine*. February 56–58, 60.

UNESCO (2015) UNESCO gives global geoparks a new label. Available at: http://www.unesco.org/new/en/media-services/single view/news/unesco_gives_global_geoparks_a_new_label/ (accessed 5 January 2021).

UNESCO (2021) UNESCO and Global Geoparks Network GGN assists geopark initiatives in Africa and the Arab states. Available at: https://en.unesco.org/news/unesco-and-global-geoparks-network-ggn-assists-geopark-initiatives-africa-and-arab-states (accessed 2 April 2022).

UNESCO (2022a) *The Operational Guidelines for UNESCO Global Geoparks*. Available at: http://www.unesco.org/new/fileadmin/MULTIMEDIA/HQ/SC/pdf/IGGP_UGG_Statutes_Guidelines_EN.pdf (accessed 20 April 2022).

UNESCO (2022b) How to revalidate, extend and rename your geopark. Available at: https://en.unesco.org/global-geoparks/revalidate-extend (accessed 20 April 2022).

White, S. and Wakelin-king, G.A. (2014) Earth sciences comparative matrix: a comparative method for geoheritage assessment. *Geographical Research* 52(2), 168–181. DOI: 10.1111/1745-5871.12062.

Yolal, M. (2012) Geotourism and geoparks: the case of Kizilcahamam Camlidere Geopark. *Geojournal of Tourism and Geosites* 10(2012), 193–203.

Zamani Farahani, H., Carboni, M., Perelli, C. and Farsani, N.T. (2019) Islamic tourism in the Middle East. In: Timothy, D. (ed.) *Routledge Handbook on Tourism in the Middle East and North Africa*. Routledge, London, pp. 125–136. DOI: 10.4324/9781315624525.

Conclusion

Anna Farmaki[1]* and Dimitrios Stylidis[2]
[1]Cyprus University of Technology, Limassol, Cyprus; [2]University of Crete, Gallos, Rethymno, Crete, Greece

The Middle East is a region that has always attracted tourism interest. Spanning a long history, the Middle East has an abundance of heritage resources and beautiful, diverse natural environments. In recent years, the area has seen rapid development primarily in terms of luxury tourism facilities in an effort to attract a more upscale tourist clientele (Villar, 2021). At the same time, the Middle East has seen its fair share of turbulence in the form of political and social unrest (Buda, 2016; Karimi *et al.*, 2021) that, nonetheless, did not compromise tourist demand long term. In fact, tourist arrivals to the Middle East had been steadily rising prior the COVID-19 pandemic (WTO, 2019). The region thus offers an interesting context for the study of tourism planning and development, not only due to the diversity characterizing the Middle East, but also because of the ever-changing tourism landscape of the region, especially in the post-COVID-19 era (Fontanari and Traskevich, 2021; Gupta *et al.*, 2022; Mimaki and Darma, 2022).

This edited volume aimed to discuss important aspects related to tourism planning and development in Middle Eastern destinations through a collection of nine chapters that draw from different study contexts. The book, in particular, presented chapters focusing on: Cyprus, Iran, Israel, Jordan, Oman, Palestine, Saudi Arabia and Turkey. While there are some papers published for individual Middle-East countries (see, for example, Öztürk *et al.*, 2013; Farmaki, 2014; Upadhya, 2016), there are limited book collections exploring various themes on tourism development and planning in the region. Specifically, this edited volume addressed varied planning and development issues ranging from policy making and sustainable tourism development to community-based tourism, heritage conservation and labour issues; therefore, bringing to the forefront of the debate on tourism impacts the potential challenges and opportunities underpinning planning and development in this subregion of the world, especially as the region is one of the most affected by COVID-19.

The book begins with a chapter on Cyprus written by Alexis Saveriades. This chapter discusses the evolution of tourism planning strategies at the destination through a critical stance with the aim of uncovering persistent governance issues that interfere with tourism developmental aspects. The author concluded that the main goals of the tourism plans, especially those designed after the Turkish invasion, paid greater attention to increased tourist numbers, mass tourism and higher revenues than to implementing the sustainability principles. The author attempts to foresee the future of the destination through its past and endeavours to answer whether Cyprus's tourism success

*Corresponding author: anna.farmaki@cut.ac.cy

 Tourism Planning and Development in the Middle East (eds S. Kladou *et al.*)
DOI: 10.1079/9781789249149.000c

story is attributed to noteworthy tourism strategic plans and exemplary governance or has it been built on false pretences and positive conjunctures.

Elad Almog is the author of Chapter 2, which examines the challenges of establishing tourism routes in Israel. Drawing on the viewpoint of public-sector stakeholders and non-profit organizations, the author discusses how disparate statuses of different areas in the routes negatively impact synergy potential and cooperation in the area. Recommendations are offered that may help overcome such obstacles. These include the following: (i) in order to connect different portions of the route to create a greater whole, each must be considered separately through the lenses of branding, scale of activity and standards; (ii) when in the process of building a branding strategy for one region alone, joining forces with another region for the creation of a shared tourism route brand becomes extremely challenging; (iii) it is proposed to establish a clear understanding of how each part of the route not only competes with but also completes the others, even when working on different scales of activity; and (iv) there is a need to clarify how a difference in standards resulting from different scales of activity might be bridged using training and certification schemes.

Chapter 3, provided by Rami K. Isaac and Wisam Abbasi, represents an exploratory study of women's empowerment in Palestine. The chapter examines the factors leading to women's empowerment through participation in community-based tourism initiatives in Bethlehem, and identifies the challenges inhibiting women's participation in such initiatives. Drawing on the perspectives of various stakeholder groups including private-sector stakeholders, the government, NGOs and local community stakeholders, the study recognizes culture and traditions, women's social status and Israeli occupation as key inhibiting factors of Palestinian women's participation in community-based tourism projects. The chapters concludes that mechanisms that facilitate the involvement of lower-educated, economically disadvantaged women and minorities will need to be further explored.

Chapter 4 by Sarasadat Makian, Ashkan Borouj and Farid Hanifezadeh, discusses how community-based lodges (CBLs) in rural areas of Iran can contribute to sustainable tourism development. CBLs as small locally owned tourism businesses, can bring economic, social-cultural and environmental benefits to rural areas. Since Iran's rural areas have great tourism potential, the development of CBLs has received considerable attention in recent years. The chapter proposes that developing new tourism products based on rural lifestyles, culture and customs can lead to the development of CBLs as a part of a community-based tourism (CBT) project. CBT has the potential to address several of the issues that rural and peri-urban areas face. Hence, the findings of this study can provide important insights for other similar destinations.

Fatemeh Bagheri, Ali Hajinejad and Naser Abdi provide Chapter 5, which focuses on heritage conservation in Iran. In this chapter, which draws from online content to analyse the views of tourism officials and authorities on heritage conservation challenges in Iran, the barriers to heritage conservation are identified and discussed, providing an explanation of the implications for the promotion of cultural tourism in the area and across the Middle East. Among them, budget shortfall was recognized as a major issue for heritage conservation projects, followed by manager attitudes towards heritage conservation in terms of contemporary urban or rural development, poor coordination among those involved in heritage conservation, poor public awareness of the importance of heritage, and legal gaps and conflicts, especially regarding the reuse of historic buildings (old houses and caravansaries) by private investors to run tourism businesses.

Chapter 6, by Selenay Ata and Sina Kuzuoglu, analyses residents' subjective well-being in relation to tourism management. Drawing on 178 surveys collected in 4 districts of the Cappadocia region in Turkey, the authors present residents' perceptions of heritage conservation and environmental preservation strategies as well as governance aspects revealing a disconnection between residents and decision-making processes. More specifically, residents have a favourable view of environmental characteristics in the region, which increases with age and is positively associated with marital status, similar to residents' satisfaction with conservation. The findings further suggest education to be an important determinant of residents' perceptions of cultural heritage. The authors conclude that Cappadocia's growth-oriented tourism strategy, with a deliberate exclusion of local stakeholders, might also result in diminishing community well-being.

Chapter 7, by Nurşah Ayhan and Stella Kladou, focuses in the Bozcaada region in Turkey to examine identity-based approaches pertaining to place branding and sustainable development. Data were collected via 16 semi-structured interviews with wine producers, local business owners, local residents and the local authority. For locals, agriculture, wine, viticulture and viniculture are part and parcel of the local culture. Wine has a crucial place not only in local people's everyday lives but also for the visitors who come to experience the island's wine culture, taste the wines and participate in wine-related activities. The chapter further discusses how this region's initial tourism development initiatives focusing on wine tourism were replaced in priority by sea and sun tourism development goals, yielding inconsistencies between local identity, place brand and local culture.

Chapter 8, by Ronen Shay, deals with multicultural tourism workforce issues. Focusing on the Eilat project, which allows Jordanians to cross the border to work in the hospitality industry in Eilat (Israel), the author examines relevant opportunities and shortcomings influencing human encounters within a tourism setting. Using data collected from 20 face-to-face interviews with various stakeholders including human resource managers, hotel managers, housekeeping managers, investors and decision makers (politicians, legislators), the emphasis was on uncovering stances and attitudes towards integrating the Jordanian workers, along with a comprehension of the effects of economic, cultural, political and security factors on the success or failure of this endeavour. Key issues identified included language barriers, historical tensions, security issues, bureaucracy, religion, ethnicity and political affiliation. The author concluded that cross-cultural synergies and intercultural management are necessary for a global multicultural workforce project.

In the final chapter (Chapter 9), Bahram Nekouie Sadry, Kholoud Mohamed Abdel Maksoud and Somayeh Zahabnazouri present the geotourism development potential in Saudi Arabia, Oman, Jordan and Iran. The development of geotourism via the establishment of geoparks is evident in the region, with an increasing attention also to its economic, conservational and educational values for the local societies. The authors highlight the importance of geotourism for local communities, primarily in terms of alleviating poverty, and present related opportunities and challenges that may inform policy makers in other regions in the Middle East. They further propose the development of a Middle East Geoparks Network (MGN) aiming to identify, preserve, manage and promote the region's rich geodiversity. Through such actions the MGN could raise awareness among local populations and decision makers about the importance of sustainable management and development of geoheritage and geotourism.

Overall, some conclusions can be drawn from the examination of the nine cases about tourism planning and development in the Middle East region. First, culture and heritage are key assets and pull factors that drive and support tourism growth and development. Second, there are many barriers that hinder tourism planning and sustainable development including lack of expertise, lack of cooperation among stakeholders and limited resident empowerment. Third, an unstable political environment is still pertinent in some places, limiting demand for such destinations. Fourth, there are unique opportunities for new tourism products based on rural lifestyles, culture and customs or geotourism. In this respect, it is evident that tourism planning and development of the Middle East needs to be elevated in importance if problems and challenges are to be tackled and opportunities are to be exploited.

The need for tourism planning and development reconsiderations is greater than ever in the post-pandemic era. COVID-19 has negatively affected global tourism by bringing international travel to a standstill (Farmaki *et al.*, 2020), forcing companies to restructure their operations and processes. Adaptations in planning and developmental processes are also required at the destination level, particularly in a flourishing region such as the Middle East where great potential as well as problems exists. As such, destination planners and policy makers in the region need to consider the increasing important of safety and hygiene as part of destination image, the changing needs of tourist markets as well as issues related to human resources.

Many issues presented in the Middle East seem quite distinct from those highlighted in other regions by: (i) Monterrubio *et al.* (2020) in Latin America, in which key themes included tourism dispersal, conflicts, unequal development, and the underrepresentation of various ethnic groups in tourism decision making; (ii) Stylidis and Seetanah (2021) in South Asia, whereby terrorism and crime, insufficient accessibility, weak institutional frameworks and overdependence on tourism, along with fragile ecosystems were in the spotlight; and (iii) Janta *et al.* (2022) in Eastern Europe, who documented themes such as: liberalization in people's movement and transport; a slow shift in peoples' and organizations' mentality still linked to the Soviet era; the prominence of cultural tourism; and a clear focus on sustainability in national and regional policy frameworks. However, there are some similarities with the themes identified by Andriotis *et al.* (2022) in Western Europe, which included the value of direct community participation in the planning process at all levels, along with the need to preserve and sustainably develop the unique cultural heritage and industrial activities that are part of the identity and image of many European destinations. All these reinforce the need for continuous research and understanding of the unique conditions and characteristics of tourism development in the various sub-regions of the world.

Although this volume contributes to our understanding of tourism development and planning in the Middle East, further research in this region is needed. First, despite the wide coverage of the area with nine case studies, some countries were not considered, such as Iraq, Kuwait, Lebanon and Syria. At the same time, each of the chapters focused on some tourism planning and development issues relevant to the country under consideration, while other themes might have been neglected. Future research is needed to capture the countries not included along with other topics, including the implications of COVID-19 in less resilient destinations.

References

Andriotis, K., Cardoso, C. and Stylidis, D. (2022) *Tourism Planning and Development in Western Europe*. CAB International, Wallingford, UK.

Buda, D.M. (2016) Tourism in conflict areas: complex entanglements in Jordan. *Journal of Travel Research* 55(7), 835–846.

Farmaki, A. (2014) Satisfaction with the rural tourism product of Cyprus. *International Journal of Tourism Policy* 5(4), 249–268. DOI: 10.1504/IJTP.2014.068027.

Farmaki, A., Miguel, C., Drotarova, M.H., Aleksić, A., Časni, A.Č. *et al.* (2020) Impacts of Covid-19 on peer-to-peer accommodation platforms: host perceptions and responses. *International Journal of Hospitality Management* 91, 102663. DOI: 10.1016/j.ijhm.2020.102663.

Fontanari, M. and Traskevich, A. (2021) Consensus and diversity regarding overtourism: the Delphi-Study and derived assumptions for the post-COVID-19 time. *International Journal of Tourism Policy* 11(2), 161. DOI: 10.1504/IJTP.2021.117375.

Gupta, S., Pandiya, B. and Shukla, V. (2022) Tourists' motivation towards destination visit intention post-pandemic: scale development and validation. *International Journal of Tourism Policy* 12(2), 160–178. DOI: 10.1504/IJTP.2022.10046332.

Janta, H., Andriotis, K. and Stylidis, D. (2022) Tourism planning and development in eastern Europe. CAB International, Wallingford, UK. DOI: 10.1079/9781800620353.0000.

Karimi, M.S., Khezri, M. and Razzaghi, S. (2021) Impacts of regional conflicts on tourism in africa and the middle east: a spatial panel data approach. *Current Issues in Tourism* 25(10), 1649–1665. DOI: 10.1080/13683500.2021.1931054.

Mimaki, C.A. and Darma, G.S. (2022) Predicting post-Covid-19 tourists loyalty: will they come back and recommend? *International Journal of Tourism Policy* 12(1), 1–23. DOI: 10.1504/IJTP.2022.121941.

Monterrubio, C., Andriotis, K. and Stylidis, D. (2020) *Tourism Planning and Development in Latin America*. CAB International, Wallingford, UK. DOI: 10.1079/9781789243048.0000.

Öztürk, Y., Çöp, S. and Sani, R.A. (2013) The relationship between corporate and regional reputation management and purchasing behaviour: an application on tourists in Ankara region. *International Journal of Tourism Policy* 5(1/2), 82. DOI: 10.1504/IJTP.2013.054053.
Stylidis, D. and Seetanah, B. (2021) *Tourism Planning and Development in South Asia*. CAB International, Wallingford, UK.
Upadhya, A. (2016) Polity, policy and destination management: an exploratory study of political systems and management of tourism with special reference to the GCC countries. *International Journal of Tourism Policy* 6(3/4), 341. DOI: 10.1504/IJTP.2016.081530.
Villar, D.F. (2021) The expansion of travel and tourism in the Middle East. UNWTO, Madrid: WTM Global Hub. Available at: https://hub.wtm.com/blog/destinations/the-expansion-of-travel-and-tourism-in-the-middle-east (accessed 29 March 2022).
World Tourism Organization (UNWTO) (2019) Tourism in the MENA Region. UNWTO, Madrid. DOI: 10.18111/9789284420896.

Index

Note: The page numbers in *italics* and **bold** represent figures and tables respectively

CABI Regional Tourism Series

BY

Konstantinos Andriotis, Middlesex University London

and

Dimitrios Stylidis, Middlesex University London

Tourism literature includes a plethora of books and conference and journal papers that address various issues of tourism in different regions and sub-regions of the world. However, up to now there has been a lack of a collection of contribution-based books with a regional and sub-regional focus. In order to complement the existing tourism literature with new or emerging topics of particular interest, it is the aim of the **CABI Regional Tourism Series** to offer a collection of books with comparative and individual case studies from at least eight regions and sub-regions of the world. These books will come under the following titles, each of which focuses on some of the countries or country clusters: Tourism in Western Europe; Tourism in Eastern Europe; Tourism in East and Southeast Asia; Tourism in Middle East; Tourism in Sub-Saharan Africa; Tourism in the Caribbean; Tourism in Latin America; Tourism in Pacific Oceania.

Suggested topics:

- Tourism Development and Planning
- Tourism Economics and Finance
- Tourism Marketing, Management and Information Technology

To discuss a proposal for a book in this series, please contact Konstantinos Andriotis at k.andriotis@mdx.ac.uk or Dimitrios Stylidis at d.stylidis@mdx.ac.uk.

Brief summary of careers

Konstantinos Andriotis is a Professor in Tourism at Middlesex University London and in the past he was Associate Professor and Head of the Hotel and Tourism Management Department at the Cyprus University of Technology. He holds an MSc in International Hospitality Management (University of Strathclyde, 1995), a PhD in Tourism Development and Planning (Bournemouth University, 2000) and a Postdoc in Tourism Marketing (Hellenic Open University, 2006). He edits the *International Journal of Tourism Policy* and the *Journal of Qualitative Research in Tourism*. He is a member of the Editorial Board of more than 20 journals and Associate Editor of the leading tourism journal *Annals of Tourism Research*. He is author of seven books and over 45 peer-reviewed articles and book chapters. His work has appeared in leading academic journals, such as

Annals of Tourism Research, Journal of Travel Research, Tourism Management and *Journal of Sustainable Tourism*. He has an h-index of 29 and more than 3700 citations. He has acted as Chairman and member of the organizing and scientific committees of various international conferences and he has given keynote speeches in several conferences. His research interests are in tourism development and planning, alternative forms of tourism, tourism experience and degrowth.

Dimitrios Stylidis is a senior Lecturer in Tourism at Middlesex University London. Prior to joining Middlesex, Dimitrios worked for 2 years as a Postdoc Researcher at Ben-Gurion University of the Negev, Israel. He holds a PhD and an MSc in Tourism Management from the University of Surrey, UK. Dimitrios is the research notes editor of the *International Journal of Tourism Policy*. His work has appeared in leading academic journals including the *Annals of Tourism Research, Journal of Travel Research* and *Tourism Management*. His current research interests include place image, tourism impacts and residents' attitudes toward tourism development.

Printed and bound by CPI Group (UK) Ltd, Croydon, CR0 4YY

13/09/2026

14961109-0007